The Big Book of Maker Camp Projects

The Big Book of Maker Camp Projects

Sandy Roberts

New York Chicago San Francisco Athens London Madrid
Mexico City Milan New Delhi Singapore Sydney Toronto

Library of Congress Control Number: 2019937652

McGraw-Hill Education books are available at special quantity discounts to use as premiums and sales promotions or for use in corporate training programs. To contact a representative, please visit the Contact Us page at www.mhprofessional.com.

The Big Book of Maker Camp Projects

1 2 3 4 5 6 7 8 9 LOV 24 23 22 21 20 19

ISBN 978-1-260-13549-7
MHID 1-260-13549-7

This book is printed on acid-free paper.

Sponsoring Editor
Lara Zoble

Editorial Supervisor
Donna M. Martone

Production Supervisor
Lynn M. Messina

Acquisitions Coordinator
Elizabeth Houde

Project Manager
Patricia Wallenburg, TypeWriting

Copy Editor
James Madru

Proofreader
Claire Splan

Indexer
Claire Splan

Art Director, Cover
Jeff Weeks

Composition
TypeWriting

To my husband, Steve; my daughters, Caitie end Gwen; my sister, Jen; and my parents, Grace and Steve: Thank you for believing in me and what I do. Thank you picking me up when I fall down and cheering me on when I succeed. No one does it alone, and I certainly couldn't have done this without you.

About the Author

SANDY ROBERTS earned her B.S. in Chemical Biology from Stevens Institute of Technology with a Minor in Literature and Certificates in Chemical Engineering and Computer Science. She did graduate work with Stevens, William Paterson, and Montana State. After working in both publishing and pharmaceutical research for several years, she had her two daughters. It changed her life.

After going back to school and changing careers, Roberts became a middle school science, math, and engineering teacher. Flipping classrooms, pushing boundaries, and helping students discover their love for STEM was too much fun to confine to school hours, so she founded Kaleidoscope Enrichment, LLC, to offer after school, homeschooling, and other local programs.

Around the same time, Roberts attended her first Maker Faire in New York City. Inspired by the maker movement, she led her first Maker Camp that summer as one of the first affiliates for the new program. In the years since, her passion for Maker Camp and for sharing the maker mindset with students has only grown, as have her programs throughout northwest New Jersey.

Roberts regularly brings her Maker Camp projects to Maker Faire and has earned multiple ribbons for her team's work. She's presented at Maker Faire, sharing her Maker Camp success tips with fellow educators. Her camp program has been featured on Makezine.com, and she shared her experiences on the Maker Camp Campfire webcasts in 2017.

Contents

1 **An Introduction to Maker Camp** **1**
What Is a Maker Camp? 1
Making It Official 3
Planning Projects and Budgeting 4
Safety 6
Keeping It Campy 7

2 **The Camp T-Shirt** **9**
PROJECT 1
Creating a Stencil with a Digital Cutter 11
PROJECT 2
Color-Changing Screen Printing 15
PROJECT 3
Permanent Marker "Tie Dye" 22
PROJECT 4
Spray-Painted Shirts 23
PROJECT 5
Bleached Shirts 25
PROJECT 6
Spin-Art T-Shirts 29
PROJECT 7
Dirt Shirts 32

3 **Camp Wearables** **37**
PROJECT 1
Ultraviolet (UV) Color-Changing Sun Bracelet 38
PROJECT 2
Simple LED Crafts 40
PROJECT 3
Light-up Proximity Friendship Necklace 42

PROJECT 4
Stamped Leather Bracelet 48
PROJECT 5
Light-up Sunglasses 50
PROJECT 6
Cosplay at Camp 55
PROJECT 7
3D-Printed Backpack Tags 62

4 Fun and Games 67
PROJECT 1
Lego Labyrinths 68
PROJECT 2
Camper "Guess Who?" Board Game 69
PROJECT 3
Green-Screened Camp Photos and Videos 74
PROJECT 4
Light-up Letter Home 77
PROJECT 5
"Smack a S'more" 84
PROJECT 6
Makey Makey Dance Off 91
PROJECT 7
Nerf Wars 101
PROJECT 8
PVC Marshmallow Shooter 112
PROJECT 9
Marshmallow Poppers 113
PROJECT 10
Creating Resin Action Figures 114
PROJECT 11
Nerdy Derby 118

5 The Campfire 121
PROJECT 1
Color-Changing Flames 122
PROJECT 2
Roasting Spy Marshmallows 124
PROJECT 3
Building a LED Campfire 126

PROJECT 4
LED Fireflies and Other Origami Fun 135

PROJECT 5
Spooky Ghost Photos 150

PROJECT 6
Making Musical Instruments 152

PROJECT 7
Music with the Makey Makey 157

PROJECT 8
Coding a Robot Drum Circle 159

PROJECT 9
Making Up a Maker Camp Song 164

PROJECT 10
Mixing Music with the NeoTrellis 166

6 Camp Food 173

PROJECT 1
Making Marshmallows 174

PROJECT 2
Making Chocolate 177

PROJECT 3
3D-Printed Cookie Cutters 187

PROJECT 4
Solar Ovens 203

PROJECT 5
Edible Treehouse 205

PROJECT 6
Campfire Popcorn 207

PROJECT 7
Sorbet Toss Game 210

7 The Great Outdoors 213

PROJECT 1
Weather Stations 214

PROJECT 2
Citizen Science at Camp 223

PROJECT 3
PVC Stomp Rocket Launcher 225

PROJECT 4
Bubbles 229

PROJECT 5
Smashed Flower Scavenger Hunt 243
PROJECT 6
Seed Bombs 245
Project 7
Sun Prints. 247
PROJECT 8
Pinhole Cameras 249
PROJECT 9
Live Model Stop Motion. 251
PROJECT 10
Making an Outdoor Game 253
PROJECT 11
Maker Rocks 258

8 Making Your Community 261

Your Maker Camp Scrapbook 261
Digital Portfolios 262
Sharing Your Maker Camp 264

Index. 267

CHAPTER 1

An Introduction to Maker Camp

What Is a Maker Camp?

The motto of Maker Camp is "Explore-Make-Share." It's an opportunity to introduce young makers and potential makers to a wonderful world of tinkering, building, inventing, crafting, engineering, and making cool stuff. It's the chance to engage kids with the do-it-yourself (DIY) maker mind-set, where anything is possible and any challenge can be overcome. It's about the belief that collaboration, creativity, and community can change the world for the better. It's a place where even the most wild, weird, and whimsical ideas can become reality.

Making things is not a new idea. Humans are makers by their very nature. Humans want to

FIGURE 1.1

understand their world and shape it. Humans love to learn. And kids, well, they still hold such wonder inside of them. They have a fundamental drive to explore and experiment.

So, if we already know how to make things, why do we need Maker Camp? Because we live in a time that sees failure as a negative experience as opposed to an opportunity to learn. Because we have systems that tend to reduce us all to statistics rather than honor the unique individuals we are. Because right now many people are more likely to buy an item than they are to make something new or fix something old. Because innovation is vitally important to making all our lives richer, better, and healthier. Because we will need new solutions to the problems we face as a global community.

The Maker Movement draws on a wide and diverse range of everything from traditional crafts to advanced technology to provide a place for new connections to form. Makers prize curiosity, determination, and grit. The Maker Movement should always be a place where everyone is welcomed, respected, and valued.

Maker Camp is the place for kids to experience this dynamic so that they can become the next generation of amazing creators pushing boundaries and making new things. Our young makers need a place to hone their skills and develop confidence. They need experiences that will make them problem solvers and self-directed lifelong learners. Maker Camp gives kids the opportunity to do this without the constraints of the classroom. Maker Camp should be fun, lighthearted, and exciting.

Each Maker Camp is different because every maker is different. Anyone can run a Maker Camp if they are willing. All it takes is the desire to learn and create with kids. If you are passionate about making something, share it with them. Let your passion be the driving force behind your Maker Camp. Plan projects that are interesting to you, not just the kids. You will be the driving force behind your Maker Camp. Embrace that.

Maker Camp provides the freedom to make mistakes, get things wrong, and try new ideas, all without judgment or formal assessment. Not only should you make sure that your kids have the opportunity to fail, but you should also be ready to model the best ways to deal with those failures, prepare your campers for the inevitable, and support them when they become frustrated. When your day is going wrong or when a project isn't working out, take time to talk about that with your campers. Let them help you fix it. Be persistent, honest, and innovative if challenges come your way so that your campers can learn to do the same.

Want to learn more about what it means to be a maker? Check out these books from your local library:

- *Free to Make: How the Maker Movement Is Changing Our Schools, Our Jobs, and Our Minds*, by Dale Dougherty (North Atlantic Books, 2016)
- *Invent to Learn: Making, Tinkering, and Engineering in the Classroom, 2nd Edition*, by Gary Stager and Sylvia Libow Martinez (Constructing Modern Knowledge Press, 2019)
- *Zero to Maker: A Beginner's Guide to the Skills, Tools, and Ideas of the Maker Movement*, by David Lang (Maker Media, Inc, 2017)
- *The Maker Movement Manifesto: Rules for Innovation in the New World of Crafters, Hackers, and Tinkerers*, by Mark Hatch (McGraw-Hill Education, 2013)
- *Mindset: The New Psychology of Success*, by Carol S. Dweck (Random House, 2006)

- *Grit: The Power of Passion and Perseverance*, by Angela Duckworth (Scribner, 2016)
- *Drive: The Surprising Truth About What Motivates Us*, by Daniel H. Pink (Riverhead Books, 2009)

Making It Official

Though there is no requirement to officially register your camp or even call it a Maker Camp, Make Media, Inc. has offered affiliate status in the past. Affiliate status is open to libraries, schools, museums, existing summer camps, makerspaces, nonprofit organizations, Boys and Girls Clubs, 4-H groups, community centers, Boy and Girl Scouts of America, religious institutions, and even family or neighborhood groups. Once you are registered as an affiliate, your camp location will be added to a worldwide map of Maker Camps. This helps those in your community who may be interested in Maker Camp to find you and your program. Typically, Make also offers a special affiliates website with a variety of resources, including project downloads, free e-books, the Maker Camp Playbook for the year, and various graphics for use in camp and marketing.

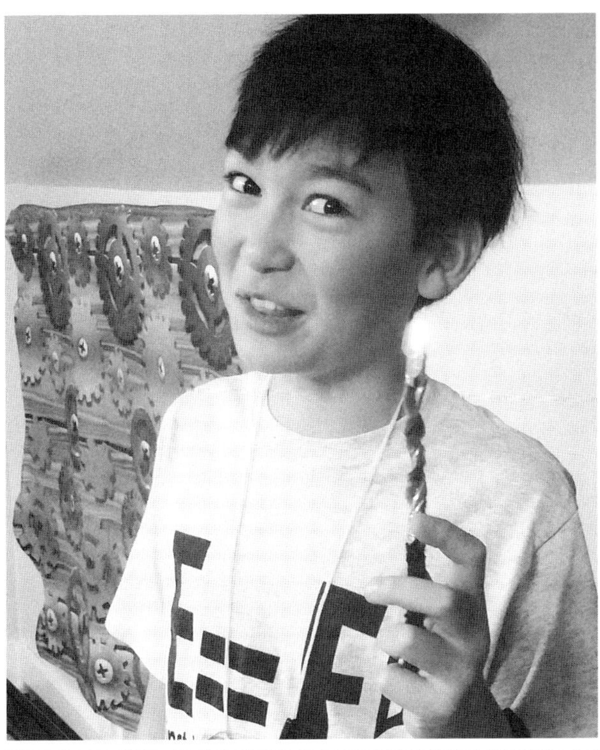

FIGURE 1.2

Make Media isn't the only maker-themed camp offered out there. For example, littleBits offers a free STEAM (Science, Technology, Engineering. Arts and Math) summer camp curriculum with projects based on their popular educational electronics sets (register at https://e.littlebits.com/summer-camp-2019). DIY.org offers hundreds of challenges for kids on a variety of topics. Participants document their work and earn badges, so it's perfect for keeping kids learning all summer. They also offer a paid Camp DIY program with over 80 projects for kids to explore. (Visit https://jam.com/courses/18 to sign up.) For younger campers, PBS offers a free Camp PBS Kids with lots of projects, outdoor challenges, and literacy connections (learn more at https://www.pbs.org/parents/page/summer). If you have kids into coding, Girls who Code (https://girlswhocode.com/), Made with Code (https://www.madewithcode.com/), and Hour of Code (https://hourofcode.com/us) all offer free curriculum that you can use and adapt to create a camp. Scratch (https://scratch.mit.edu/) even offers a special Scratch Camp with themed projects for kids to create and share.

The most important thing to remember is that you do not need to wait for any official registration to open to start planning your camp. Especially with larger organizations, you often need to have projects planned, materials and staff budgeted, and marketing out in late winter or early spring (or earlier), so it's not always practical to wait for the official announcement. Don't let that hold you back. Start planning early and draw from whatever sources look exciting to you.

Planning Projects and Budgeting

Think of planning a Maker Camp as a big, exciting maker project of your own. Consider what skills you want your campers to develop. Think about their interests, what needs there may be in your community, and what materials you already have on hand to use for projects. You can plan a camp with one overarching theme, organize ideas into project paths, or make each day its own unique adventure.

Obviously, you have this book, so you are off to a good start. You can just simply select projects that you like and start on your materials list and budget. But if you want to add additional projects to fill out your Maker Camp experience, you'll need to do some research.

FIGURE 1.3

I usually start gathering ideas in the fall by attending Maker Faires. These events provides a ton of inspiration. If you have a Maker Faire event near you in the summer or fall, consider going to see what the trends are in the maker community. Even if you can't physically attend, follow the Maker Faire on social media to see what's hot.

Of course, you'll want to check out *Make Magazine* regularly for project ideas. *HackSpace*, *Nuts & Volts*, *Circuit Cellar*, and *Popular Science* are also great places to start. Websites such as Instructables.com and DIY.org offer many projects of all types. And, of course, there are many excellent books offered by McGraw-Hill/TAB, Make Media, DK Publishing, and No Starch that offer ideas and tutorials. Check out HumbleBundle.com regularly for great deals on e-books.

Remember that Maker Camp is a chance to play, so take a look at what's big right now. What video games, movies, music, TV shows, or books can you draw from that will engage your kids? What new trends are there in crafting or technology? What crazy fads are kids interested in? If you can take something kids have been consuming and reframe it as something they can make, that can be powerful and transformational.

Once you've brainstormed with your team and selected projects for the year, you'll want to plan a schedule. Plan to host a "get to know you" style project toward the beginning of the week, such as the camper "Guess Who" board game feature in this book. Wearables are also perfect for the first day or two. Make a T-shirt or have campers hack their name tags to show off their personality. Toward the end of the camp, plan for activities such as videos, journaling, and scrapbooking.

Each day, plan for at least half an hour of settle-in time. Provide building toys and arts supplies so that the campers can get their maker brains engaged. I usually end days in the same way. Resist the urge to schedule every minute for your makers. They need time to create on their own and to get to know each other. I also like to plan a "debrief" at the end of each day so that campers can share what they learned.

Each day, make sure that you plan at least one outdoor activity, weather and location permitting. Even if you're outside to launch rockets or fly drones, take time every day to get some sunshine. Many kids don't get enough time in the great outdoors, and Maker Camp can provide a unique opportunity to get outside and get creative. Have the kids build structures out of found branches and other materials. Get inspired by nature artists such as Andy Goldsworthy. Or just plan some time to run around. Added bonus: all the blood flow, oxygen, and endorphins are fantastic for creativity!

Start your budgeting early if you plan to use equipment such as microcontrollers (micro:bits, Circuit Playground, Arduino boards, Makey Makeys, etc.). In January, look for potential grants to help fund these "big ticket" items if needed. If you are unable to raise funds, reach out to local makerspaces, schools, and colleges to see if there are items you can borrow.

I usually try to coordinate projects that use similar materials, and in many cases, I did the same with projects in this book. If possible, coordinate with others and order in bulk from such places as S&S and Oriental Trading Company. You may be able to get some items, such as drinking straws, bamboo skewers, and cardboard, donated from local grocery stores. Reach out to the local hardware stores and craft stores as well. They may be willing to donate or discount items for your camp in return for a little promotion. The truth is, it never hurts to ask.

Your cost will vary by the length of your program, the number of campers you plan to host, the complexity of your activities, and the amount of staff needed. For something like my Science and Swim Maker Camp, my materials are usually recycled and upcycled water bottles and the like. My only real expense tends to be colorful duct tape and bubble mix. So I usually budget about 50 cents per camper per day. Library programs usually run 60 to 90 minutes with about 25 campers. I keep that cost at $1 to $2 per camper per day. Full week programs tend to have more advanced projects, so I may budget as much as $5 per camper per day for that. Create a spreadsheet of your materials and costs. Doing this each year will help you to better track your spending and make budgeting easier over time.

You'll likely need staff as well. In many cases, you can find volunteers to help. Parents are a great choice, especially if you're working with 4-H, scouts, a church, or similar organizations. You can also often find teens and college students that need or want community service hours who are great for Maker Camp. Teens can often be more approachable for campers, and there can be less fear of judgment when a project fails simply because the age difference is minimal. Use this to your advantage. Since I've been hosting Maker Camps for years, I like to bring back former campers. Don't count out retirees either; you'd be amazed at the wealth of knowledge they have and are willing to share with your young makers.

As we come up to Maker Camp each year, I create an outline for each day in Google Docs. The outline will include the projects, the approximate time each will take, and a link or two to information about the project. I share this with my staff so that they can familiarize themselves with what we'll be doing in advance.

If you are able, host a Making Maker Camp meet-up and have everyone try out the projects. This is a great way to find any potential challenges and get great input from your staff.

Safety

At Maker Camp, you are inherently going to run into safety issues. You're using tools and electricity, after all. But you don't need to fear challenging projects. Just plan ahead.

Make sure that you have good-quality eye wear that will fit your campers. Buy actual safety goggles, not costume items. You want to make sure that they won't shatter. Most campers prefer safety glasses to safety goggles because goggles can fog up and can be uncomfortable. Unless you are doing a project that really needs the goggles, go with glasses. Polycarbonate lenses and frames are the most common. Spend a little more if you must on scratch-resistance glasses. You may want to assign a pair of glasses to each camper. In addition, make sure that the glasses are washed regularly in warm, soapy water. Let them dry, and then remove spots with a microfiber cloth.

Gloves are another must-have. I generally use three types in my camps: powder-free nitrile disposable gloves, knit work gloves with rubber or nitrile coating, and leather work gloves. Disposable gloves are great for foods and crafts. I don't use powered gloves or latex because they can cause allergic reactions. The knit gloves are great for lightweight projects such as sanding or using hot glue. The work gloves I consider mandatory for any kind of cutting activity. You'll want to order small, medium-sized, and large gloves of each type. It's an investment. This is a great place to ask a local hardware store for help. For both the knit and the leather work gloves, I use clothespins to keep the left and right gloves together.

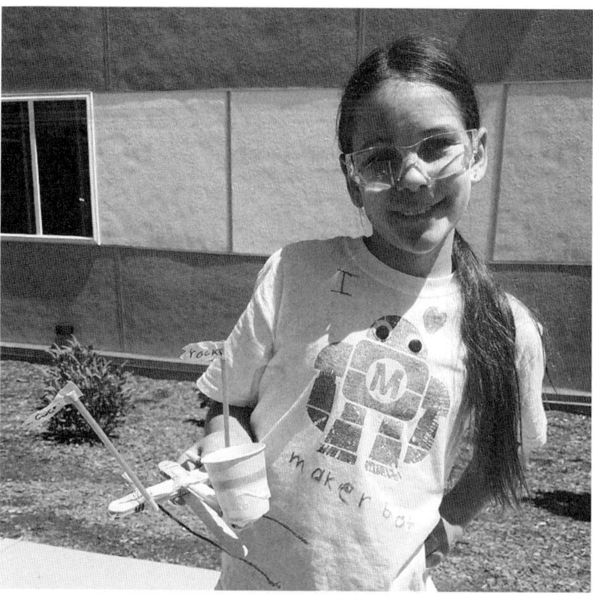

FIGURE 1.4

You may also need ear plugs or ear protection depending on your projects. It never hurts to have dust masks on hand either. I usually keep a pair of heat-resistant silicone gloves or mitts on hand too.

Campers should arrive at Maker Camp in clothes and shoes that are appropriate for your projects. This usually means no loose-fitting clothing, close-toed shoes, or sneakers, and campers should keep long hair tied back and remove jewelry as needed. Because you will be meeting in summer, you may want to have safety aprons to protect campers' legs if they wear shorts.

Make it a point to go over basic safety rules as the first thing you talk about on the first day of camp. Rules may include not running or horsing around while tools are in use, always working with a buddy, wearing safety equipment, being trained on tools before using them, and other basics. Ask your campers to contribute to the list. Have a safety policy, post it, and review it every day. Make sure that all campers are committed to keeping one another safe. The

Makerspce Playbook: School Edition has a great chapter on safety if you want further ideas.

Often, over the summer, even if you are in a school, you won't have a nurse on hand for injuries. I recommend that you always have someone with first aid and cardiopulmonary resuscitation (CPR) training at camp. Courses are often offered through local hospitals, so plan ahead and get certified. Have a good first aid kit on hand as well. You can either assemble one yourself or buy a complete kit. Other safety equipment to have on hand includes a fire extinguisher and/or a fire blanket.

It also helps to take a moment to make a safety plan. Make sure that campers know what to do if a fire alarm goes off. If there is an accident, how should they alert you, and what should they do? Although you may not want to, you should go over lock-down and active-shooter protocols as well.

While you're discussing safety, you may also want to have a discussion about your code of conduct. I always post one. Take time to collaborate with campers about what a respectful, supportive, exciting Maker Camp looks like. I often use phrases such as "Process over product," "Stay positive and productive," and "Create, don't consume" to remind campers what Maker Camp is all about. It should be made very clear to campers that harassing, insulting, and intimidating others will not be tolerated. And they should be reminded to be respectful of the opinions and abilities of others. Lastly, encourage campers to create teams that are open, welcoming, and supportive of each member and that find ways for everyone to contribute.

Keeping It Campy

My last bit of advice—and this is important—is to keep it campy. Keep it fun. Keep it creative. Keep it positive. Keep it hands-on and DIY. Keep it camper focused and camper led.

At the heart of the maker mind-set is a willingness to push boundaries, experiment, hack, play, and try something new. So, while you want to have safety rules, an organized space, and a solid schedule, you always want to allow space for flexibility and adaptability. Embrace the freedom of summer!

Plan goofy camp traditions. Make camp T-shirts. Make s'mores! Come up with special songs, camp mottos, or camp greetings. Whatever you do, don't let your Maker Camp become another boring class or workshop. Summer camps should be fun, and Maker Camp is no exception.

FIGURE 1.5

CHAPTER 2

The Camp T-Shirt

There is something special about being a member of a like-minded, supportive group. This is why we make T-shirts for special occasions, sports teams, school recognition, business promotion, and, of course, summer camp. We want to say, proudly, I'm part of something bigger than I am. It's no different with the Maker Movement.

This is why my first project for Maker Camp every year is to make a camp T-shirt (Figure 2.1). Rather than having T-shirts printed up in advance, we go DIY

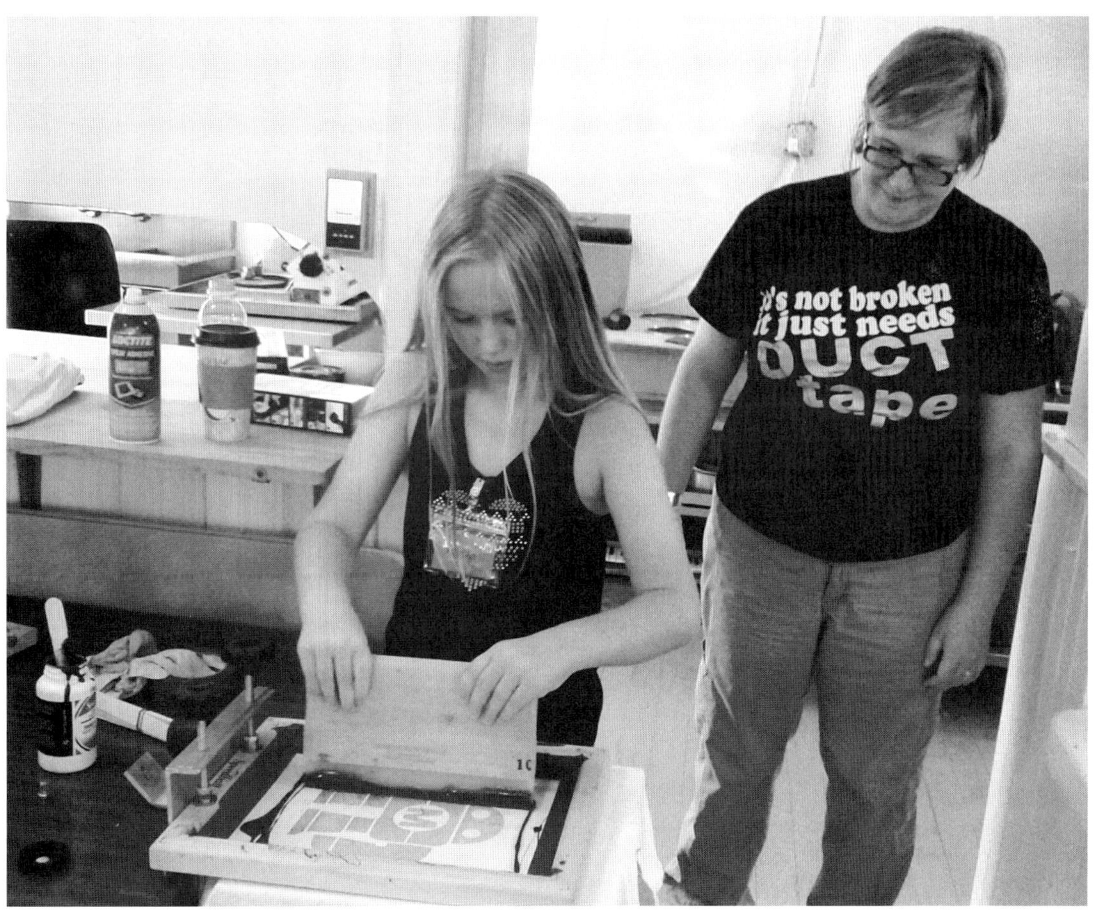

FIGURE 2.1

on this project, just as we do with everything else. Each year my campers and I explore a new T-shirt-making technique. As anyone who has ever visited Pinterest knows, there are a lot of ideas out there for customized shirts, some better than others.

Regardless of the method, there is one thing I've kept the same each year: the logo. The Makey robot is a symbol of the Maker Movement and is essentially the adorable metal mascot for Maker Camp. So each year he adorns the front of our camp T-shirts. This gives campers lots of room on the back to add their own designs, try a second technique, or simply sign one another's shirts as a keepsake.

PROJECT 1
Creating a Stencil with a Digital Cutter

COST: $–$$$

MAKE TIME: 15–30 minutes

Unless you have a particularly skilled group of artists in your camp, you will need to make a stencil to add Makey to your shirts. There are two ways to do this: (1) a traditional hand-cut stencil or (2) a new-fangled digitally cut stencil. Personally, because I often need quite a few stencils, I like to cut them digitally. It's surprisingly easy to do and takes considerably less time than cutting a stencil with an X-ACTO knife.

In this case, I'll be using a popular, low-cost digital cutter called the Silhouette Cameo and its free software, Silhouette Studio. There are many similar digital cutters on the market now, each with its own software, but most work in very similar ways. You may need to refer to your individual cutter's user's manual for specifics.

Once you've got your stencil designed, you then have lots of options about what to use as the stencil medium. If you want to do classic screen printing, you'll want to cut your design from adhesive-backed vinyl and attach it to a fabric screen with a wooden frame. You can use plastic stencil film sprayed with adhesive for wet applications of spray fabric ink or bleach. Cardstock with spray adhesive is an inexpensive solution for single-use stencils or when using markers, spray paint, or "dry brushed" acrylic paints. Cutting freezer paper allows you to iron on stencils for single use with almost any application.

Whatever your final application, it all starts with creating and cutting the design.

Materials

- A computer with access to the Internet
- A digital craft cutter, such as a Silhouette Cameo
- Digital cutting software
- Medium for the stencil: adhesive-backed vinyl, freezer paper, cardstock, or stencil film
- Weeding hook
- Spatula (or old credit card), if needed to lift stencil
- Fabric ink or acrylic paint that is appropriate for your project

From Pixels to Digital Design

1. If you don't already have the image you want, start by using an Internet search engine to find one. Look for solid, crisp, clear, one-color images. Get the largest sized image you can find.
2. Import your image to your digital cutting software. In Silhouette Studio, you simply have to open the graphic (Figure 2.2).

FIGURE 2.2

3. Resize your graphic. In most software, you can click on the image, select a corner, and drag the image to resize it. In Silhouette Studio, you can also select the image and directly adjust the scale using the toolbar or Scale window found under Panels > Transform. I suggest setting the width to 8 inches for starters. (For more information on sizing your image, see "Cutting Your Design," next.)
4. Use the software's Trace function to create a cutting outline. In Silhouette Studio, you can find this under Panels > Trace. Click "Select Traceable Area," and drag the box over your image. Under "Trace Style," select Trace. At this point, you can remove the original graphic. The cut lines should be left behind (Figures 2.3 and 2.4).

FIGURE 2.3

FIGURE 2.4

5. If you are planning to use this stencil as a vinyl transfer for screen printing, you can skip this step. Otherwise, you may need to add elements to connect all the pieces of your design so that it can be lifted as one piece. In technical stencil terms, you want "bridges" to connect the "islands."
6. To create bridges, use the drawing feature in your software. Simply make rectangular boxes, $\frac{1}{8}$ to $\frac{1}{4}$ inch thick, and place them between the pieces that need to connect. Then use the Subtract function (found under Panels > Modify). You can rotate your bridges by clicking on the green dot above the box or selecting Object > Rotate. When you are done, group your pieces (found under Object > Group) (Figures 2.5 and 2.6).

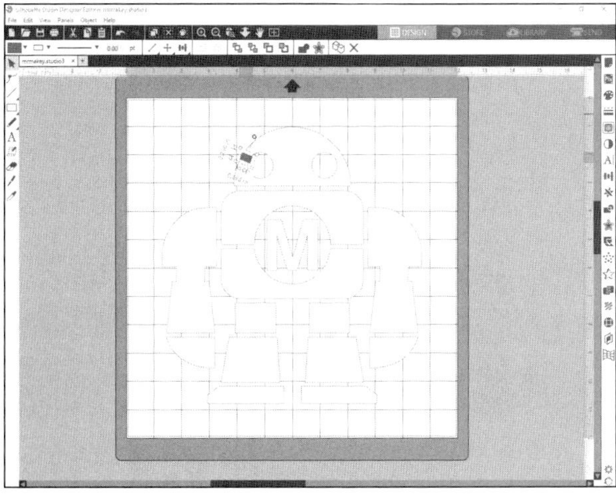

FIGURE 2.5

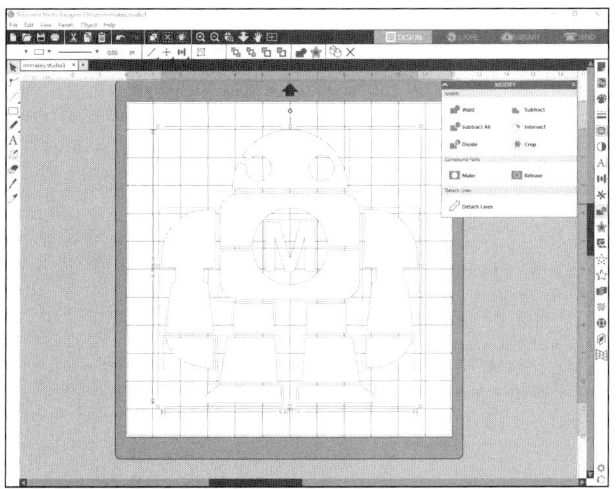

FIGURE 2.6

7. Check your stencil. Fill your grouped pieces by selecting the image, clicking on Panels > Fill Color. Then make a 10- × 12-inch rectangle and fill it with a contrasting color. Place this over your stencil image, and then right-click and select "Send to back." The contrast color of your box is what you will use as a stencil, so make sure that it is all connected (Figure 2.7).

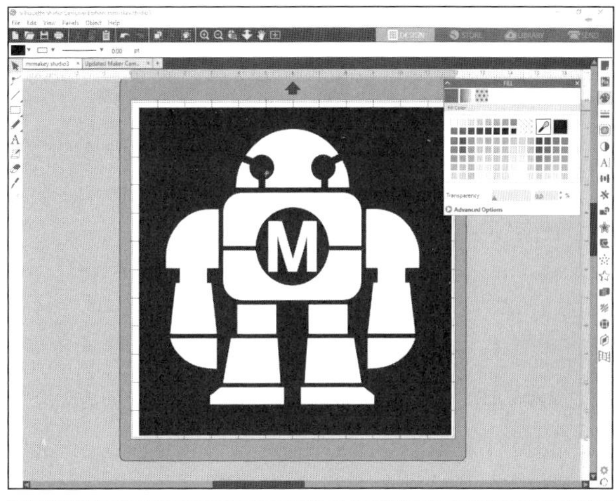

FIGURE 2.7

Cutting Your Design

1. Before you make your cut, you'll want to check the size of your image. Most T-shirt printers recommend that designs range between 8½ inches square for extra small and small sizes and 10½ inches square for medium and large sizes. A standard adult T-shirt can accommodate a design of up to 11 inches square.
2. You can cut your design at different sizes depending on your shirt sizes, but this may add a level of organization and complexity that can become challenging at camp. Personally, I stick to a design that is 8 to 10 inches square and let campers use the extra space around the design, if there is any, to customize their shirts. Adjust your design on the grid as needed to have it centered horizontally.
3. You'll want to plan to have about 3 inches between the collar and the top of your design. Although you can place your design "on the fly" by using the width of an adult hand, fingers together, no thumb, I find it can be easier to plan this distance on my cut, so I just have to match the top of the medium against the collar.

4. Attach your medium to your cutting mat. For vinyl and stencil material, the paper backing should face the mat. For freezer paper, place the medium shiny side down.
5. Adjust your cutter to the correct setting for your medium. I find that it helps to slow the cut a bit in the settings if you have sharp turns in your stencil. You may need to play with your settings to get a good cut, so don't wait until the day before you need your stencils. I speak from experience here. Seriously.
6. Once you have cut the stencil, you need to weed it. Use a weeding hook (or an X-ACTO knife) to remove the unneeded parts of the medium. Then carefully remove the stencil from your mat.

Applying the Stencil

1. Place a piece of cardboard in between the layers of the shirt. Although you can cut your own, I have found that one-quarter sheet of rectangular cake sheets work very well, are inexpensive, and are easy to buy at craft stores or online. Buying enough so that each camper has one to use will let you prep shirts while others are decorating, saving time. It will also allow you to leave the cardboard with the shirt while it dries. Plus, you can label the carboard with the camper's name rather than writing on the shirt.
2. Place the stencil over the shirt and center it. To find the center, fold the T-shirt in half vertically, matching the sleeves. You can mark it with a piece of chalk or iron the fold to make a crease. You'll want the top of your image to be about 3 inches (or the width of your hand, fingers together, thumb not included) below the collar.
3. Apply the stencil.
   a. If using cardstock or stencil material, lightly apply a coat of spray adhesive. Allow it to dry for 1 minute. Place the stencil adhesive side down, and firmly smooth it over the fabric, starting in the center and pressing out any bubbles.
   b. If using freezer paper, place the stencil shiny side down onto the shirt. Using a dry, hot iron, heat the stencil starting in the center and moving outward with a slow, circular motion until the paper is well attached. Be careful not to burn the paper (Figure 2.8).

FIGURE 2.8

   c. If using self-adhesive vinyl or stencil material, peel back the first inch or two from the top of the stencil. Center the stencil in place, and press the top to the shirt. Slowly pull away the backing, pressing the stencil to the shirt as you go.

PROJECT 2
Color-Changing Screen Printing

COST: $$

MAKE TIME: 15 minutes, more if making your own screen, plus drying time

One of my earliest day camp memories from my own childhood was learning to screen print a T-shirt. I got to make my own simple design, place the screen on the shirt with a big professional setup, and drag the fabric ink across with a squeegee. That shirt was my pride and joy for quite a while, and the experience is probably one of the reasons I'm so committed to making T-shirts with my campers each year.

This is also why the first time I held a Maker Camp, I went out and got myself a Speedball Screen Printing Kit and went to work. You know what? It was a mess.

Traditionally, screen printing involves a process of applying a chemical emulsion, applying a design reverse printed to transparent acetate, shining a bright light on the design for just the right amount of time in a dark room, scrubbing off the residual emulsion, drying the screen, and then using it. When you're done, you apply another chemical to remove the emulsion, do a lot more scrubbing, and then you can start all over again.

Suffice it to say that it's not exactly an easy process to do at home or in a classroom. I quickly realized that I needed a better process if I was going to do this every year.

Enter *Make Magazine*. In May of 2015, Chris Connors published *Simple Silk-Screen Printing Using a Vinyl Cutter* (*Make Magazine*, Maker Media, Inc.). It detailed using a digital vinyl cutter to create a mask that was then laid onto a silkscreen. Because I already had a Silhouette at the time, I gave it a shot. The rest is history.

Cutting vinyl and applying it to a traditional fabric screen is much easier, quicker, and less expensive than the traditional emulsion process. It also makes it so much simpler to change your design or reuse your screen, which means more creative opportunities for your campers!

Because you are now saving all this time, you can give your fabric inks a bit of love. Although screen printing with vinyl works with traditional fabric inks, you can mix your own as well. For Maker Camp 2016, I used thermochromatic (heat-sensitive), photochromatic (light-sensitive), and phosphorescent (glow-in-the-dark) pigments to create our camp T-shirts. It was so much fun, I brought the project to Maker Faire that year.

Materials

- Your stencil cut on adhesive-backed vinyl
- Transfer paper or tape
- A cotton T-shirt, washed and dried
- A translucent or transparent base
  - I suggest Speedball Fabric Transparent Base. It's inexpensive, water based for easy cleanup, and easy to find at many craft stores.
- Reactive dry pigments, 2 to 3 grams per 4 ounces of ink
  - I like Solar Color Dust (https://solarcolordust.com/) pigments. Plan for 2 to 3 grams of pigment per 4 ounces of base. It is sold in 10-gram quantities for around $20, which means you'll be able to make many projects. The company also offers a sample pack, with 1-gram samples of 10 colors.

- A silkscreen
  - Either a prepared silkscreen in a frame, 10 × 14 inches, 110 mesh, or
  - Your own DIY screen, which uses a wooden picture frame, 11 × 14 inches, and sheer nylon fabric or silkscreen
- Staple gun (if making your own optional silkscreen)
- Scissors
- Masking tape or blue painter's tape
- A piece of 9- × 12-inch cardboard
- Small painter's spatula or disposable spoon
- Squeegee (or an old credit card for smaller projects)

Making Your Own Silkscreen (Optional)

1. Select your fabric.
   a. You can buy silkscreen printing mesh online. This synthetic polyester fabric is sold based on its mesh, which refers to how tightly woven the fabric is. The lower the mesh, the larger the spaces between the strands of thread in the fabric, allowing for more ink to flow through the fabric. The higher the mesh, the more tightly woven is the fabric. A mesh number of 110 is a popular choice for many applications. A higher mesh can give you a finer, more detailed image and will handle more watery inks, but it takes more work to get the ink through the fabric.
   b. You can use other, more common fabrics to reduce your cost. Try fine-mesh nylons, such as those used for curtains or drapes, for example. This is a great opportunity to upcycle fabric that you are no longer using. Be sure to wash and dry your fabric beforehand. You don't want dirt, oils, starch, or sizing to block the adhesion of the vinyl to the fabric or stop the flow of your inks when printing.
2. Prepare your picture frame. Remove the glass and carboard backing from your frame. Remove or bend flat the metal tags that typically hold the backing and the picture in the frame.
3. Place the frame, front down, on the fabric, and cut around it, leaving 1 to 2 inches (depending on the thickness of your frame) around the outside (Figure 2.9).

FIGURE 2.9

4. Starting on a long side, fold the fabric over from the front to the back of the frame, straighten and staple. Don't skimp on the staples, by the way. You want the fabric held tight and strong (Figure 2.10).

FIGURE 2.10

5. Grasp the fabric on the opposite side, and pull it taught, folding it over the back of the frame as well. Starting in the center, staple the fabric to the frame. Keep the fabric held tightly as you staple along the edge. Repeat for the short sides of the frame. This is an important step. Take your time to get the fabric tight and flat.
6. Pull the corners up, make sure the fabric is tight against the frame and fold it down into a triangle shape, and then fold it against the frame. Staple. The goal here is just to keep the corner fabric tight and flush against the frame. Trim any excess fabric.

Applying the Vinyl to Your Screen

1. After cutting and weeding your vinyl stencil, cut a piece of transfer paper that is slightly larger than your stencil image. Starting at the top, attach the transfer paper to the front of the stencil, rolling it over the stencil and removing the backing as you go. Be careful not to create bubbles or damage the delicate stencil.
2. Now flip the stencil–transfer paper sandwich and remove the backing from the vinyl. Again, take care not to damage the vinyl in the process.

3. Orient the stencil over your frame, center it vertically and horizontally, and slowly roll it onto the fabric, again starting from the top and being careful to avoid bubbles or stretching. Press the vinyl stencil firmly to the screen fabric. Once the stencil is transferred, carefully remove the transfer paper. This is the most challenging part of the process, in my opinion, especially if your stencil has small details (Figure 2.11).

FIGURE 2.11

4. Use the masking tape to cover any areas left uncovered by the vinyl. Make sure that the edges along the frame are well sealed by folding the tape so that it forms an "L" between the side of the frame and the screen (Figure 2.12).

FIGURE 2.12

Mixing Your Ink

1. The suggested starting ratio of dry pigment to clear base is 2 grams to 4 ounces. If you don't have a digital scale available, approximately 1 teaspoon of dry pigment mixed into your base is a good place to start.
2. Add the pigment, mix it very well, and then test a small amount on a scrap of fabric or paper. I've found that the ultraviolet (UV)–reactive dyes need a higher ratio of pigment to base than heat-sensitive or glow-in-the-dark pigments, so testing to make sure that you have the effect you want is important. Every pigment is a little different.
3. Once you have the color you want, give the mixture another stir, and let it sit overnight. This helps the dye to saturate the base, and I've found that it gives better results.

Question: *How much ink will I need?*

This is actually not an easy question to answer. The amount of ink you'll use depends on how thick your ink is, what type of ink you are using, how fine your mesh is, how large and open your image is, what color ink you are using, how much pressure you use when printing, the material onto which you are printing, and several other factors.

Professionals have mathematical equations to help them figure out the amount of ink they'll need for a project. If you're interested in taking the deep dive (or making it a project for your students), there are several tutorials available online (e.g., http://www.screenweb.com/content/how-much-ink-do-you-need).

That said, I usually get 15 to 20 shirts printed with an 8- to 10-inch square Makey for every 4 ounces of ink. Of course, this can vary a lot when you are working with campers because some will be more naturally skilled at the printing process than others. Plan to have extra ink ready to go, just in case you need it.

Screen Printing

If you have access to a screen printing press and platen, use it. Here's the deal. The platen is a flat, mostly rectangular piece of wood onto which you put your T-shirt. That's attached to a metal vice that holds it steady on the edge of a table. Above the platen, there is another vice that holds your screen along the top of the wooden frame. Between the two, there is a hinge that allows you to pull the screen down onto the T-shirt, print, and then lift it back up. It's designed to make T-shirt printing faster. It isn't a required piece of equipment, but if you have access to one, if will make your camp experience easier (Figure 2.13).

FIGURE 2.13

But, if you don't have one . . .

1. Place a piece of cardboard in between the layers of the shirt.
2. Place the screen over the shirt, and center it. To find the center, fold the T-shirt in half vertically, matching the sleeves. You can mark it with a pencil or a piece of chalk if you like or iron the fold to make a crease. You'll want the top of your image to be about 3 inches (or the width of your hand, fingers together, thumb not included) below the collar.
3. Scoop some of your ink onto the screen edge, creating a line of ink. Hold your squeegee (or, for small projects, an old credit card) at a 30- to 45-degree angle, and drag the ink across the image, going in only one direction, usually from top to bottom. The first pass will "flood" the image and fill the tiny holes in the fabric with paint. You shouldn't need to flood the image for every shirt.
4. Reload the ink, and use your squeegee to drag the paint across the image, using pressure this time to push the ink through the fabric onto the shirt. Aim for a consistent single pull of ink across fabric. This will push the ink through the fabric where it is unmasked and onto the shirt. It's important not to load too much paint or make too many passes for a single print because the ink can bleed under the stencil.

FIGURE 2.14

5. Do a quick visual inspection to see if the image is covered with ink. Lift the screen, and check your print. Let the paint air dry or use a blow dryer to set the ink (Figure 2.14).
6. To heat set the ink onto the shirt, turn it inside-out, and put it in a dryer for 30 to 60 minutes. Don't wash the shirt for a week to allow the paint time to cure (or follow the instructions given for your specific ink).

FIGURE 2.15

NOTE: In general, be careful not to expose the inks to heats over 200°F because that can degrade the ink and cause the color change to stop happening. The color change should last an amazing 5,000 cycles before it finally gives out, so you'll have lots of opportunities to play (Figure 2.15).

PROJECT 3
Permanent Marker "Tie Dye"

COST: $

MAKE TIME: 5–15 minutes, plus drying time

This is the easiest of all the Maker Camp ideas. If you want something that doesn't cost much, has a simple setup, and offers lots of room for camper creativity, this is the way to go (Figure 2.16).

FIGURE 2.16

Materials

- Your stencil cut on plastic stencil material, cardstock, or freezer paper
- A white cotton T-shirt, washed and dried
- A piece of 9- × 12-inch cardboard
- Permanent markers in a variety of colors (must be alcohol based)
- Isopropyl alcohol, preferably 90%
- 2-ounce spray bottle
- Large plastic tub or bin
- Goggles

Process

1. Apply your stencil to the shirt as described previously.
2. Color within the stencil area using the markers. Campers do not need to fill the area in completely. Encourage campers to leave space between colors. Using repeating patterns with different colors placed near one another often yields the best results.
3. Place the shirt in the bin. Have campers put on goggles to protect the eyes. Gently spray the alcohol over the shirt. Encourage campers to not soak the shirt but rather add a little alcohol at a time to see what effects they get. The alcohol will spread the dyes in the ink, blending nearby colors.
4. Lay shirt flat to dry. The alcohol usually evaporates quickly, leaving just the color behind. Campers may want to apply more alcohol until they get the effect they like.
5. To heat-set the ink onto the shirt, turn it inside out and put it in a dryer for 30 minutes.

PROJECT 4
Spray-Painted Shirts

COST: $

MAKE TIME: 5–15 minutes, plus drying time

This is another quick, easy T-shirt-making method with a lot of flexibility in the paint you use. The simplest path to T-shirt greatness is to buy readily available fabric spray paints at your local craft store. Available in a wide array of colors including glitter, glow, and metallics, water-based fabric spray paints (such as those sold by Tulip brand) are fairly economical and offer a lot of creative opportunity. The colors blend nicely, are permanent once they dry, and are easy to apply.

In addition, aerosol versions are also available. These paints can be a bit trickier to work with, but they provide opaque coverage without blending, so colors can be layered. They can also be used on colored shirts rather than the standard white you need for water-based paints. If you're working with teens or adults, you can even use regular spray paint, although the final result may be stiffer than fabric paint, which some campers find uncomfortable to wear.

If you want to make your own spray paints, it's very easy to do so. Use pliers to open dead permanent markers and pull out the long felt inside. Place this in a 2-ounce spray bottle and cover it with isopropyl alcohol. Let stand at least 24 hours. Remove the felt, and spray as described below.

Materials

- Your stencil cut on plastic stencil material, cardstock, or freezer paper
- A cotton T-shirt, washed and dried
- A piece of 9- × 12-inch cardboard
- Spray paint of choice
- A large cardboard box
- Goggles

Process

1. Apply your stencil to the shirt as described previously.
2. Place the T-shirt in the box. Put on goggles. Holding the spray paint at least 6 inches away from the shirt, gently spray the paint over the stencil. Use more than one color as desired.

NOTE: If you are using aerosol paints, either work outside or be sure to work in a well-ventilated room with a fan for air circulation (Figure 2.17).

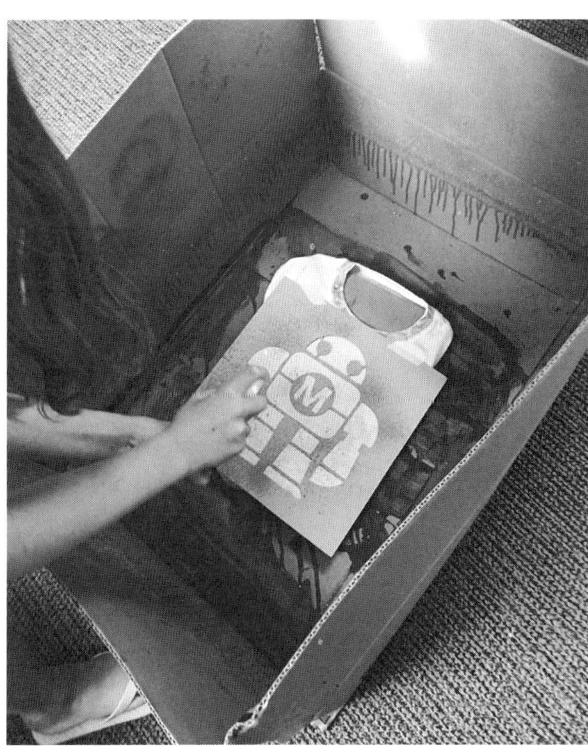

FIGURE 2.17

3. Lay the shirt flat to dry. If you are using aerosol sprays, drying time may be 30 minutes or less. Water- or alcohol-based paints will take longer to dry.
4. To heat-set the ink on the shirt, follow the directions provided by the manufacturer. If you are using standard spray paint, heat-setting is unnecessary (Figure 2.18).

FIGURE 2.18

PROJECT 5
Bleached Shirts

COST: $

MAKE TIME: 5–15 minutes, plus drying time

Using bleach on T-shirts is really fun because, unlike with other processes, you're removing color rather than adding it. This is a great activity for older campers looking for something new. (It's not a good idea with younger campers, though, for safety reasons.) Once you've removed the color of the shirt, you can use the other methods listed earlier to add new colors to the bleached areas, if desired.

Commercially available bleach pens are a great tool for this as well. They provide a good tip so that with slow and steady application, campers can add text and details to their creations. Campers should apply the bleach as they would icing on a cake, leaving a thick line that will lift the color from the shirt (Figure 2.19).

I suggest black shirts in the "Materials" list, but any darkly colored shirt should work. Because you can't really know what color will be revealed until after you apply the bleach, you may want to test your shirts first to make sure that the color lifts nicely. Applying heat, for example, using a blow dryer on the bleached area can speed up the reaction and yield light colors. Don't push it too far, however, or the bleach can eat right through your shirt!

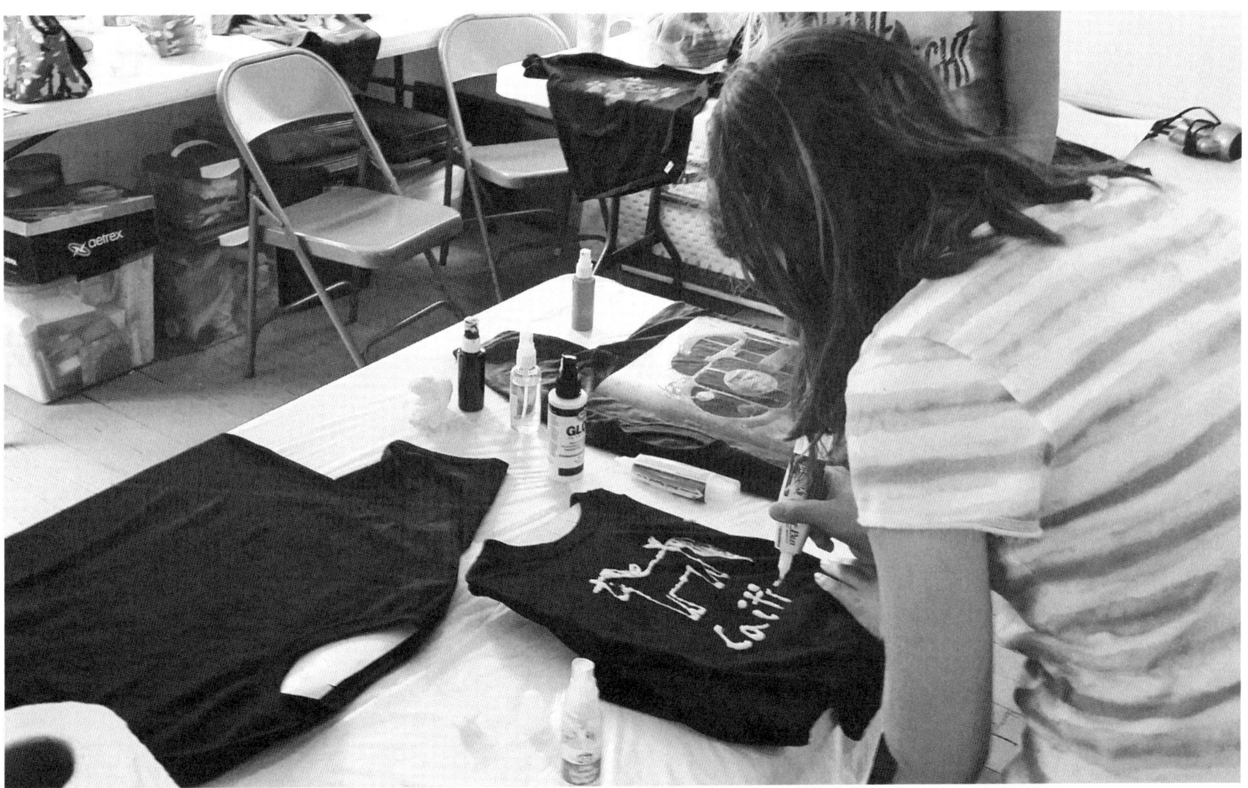

FIGURE 2.19

Materials

- Your stencil cut on plastic stencil material, cardstock, or freezer paper
- A black cotton T-shirt, washed and dried
- A piece of 9- × 12-inch cardboard
- Bleach pens (optional)
- Bleach
- Water
- 2-ounce spray bottles
- A large cardboard box
- Goggles

Process

1. Apply your stencil to the shirt as described previously. For this treatment, you may want to consider taping newspaper to mask areas surrounding the outside of the stencil (Figure 2.20).
2. Place the T-shirt in the box. Put on goggles. Holding the spray bottle of bleach at least 6 inches away from the shirt, gently spray the bleach over the stencil (Figure 2.21). It is best to spray a light coat, give it time to react, and then add more. You can use a blow dryer to heat the bleach and speed the reaction, if desired.

FIGURE 2.20

FIGURE 2.21

NOTE: When using bleach, either work outside or be sure to work in a well-ventilated room with a fan for air circulation.

3. If desired, draw on the shirt using the bleach pens.
4. Lay the shirt flat to dry. Using a blow dryer on a warm setting can speed the process.
5. Wash and dry the shirt as soon as possible to remove residual bleach (Figure 2.22); otherwise, holes may form (Figure 2.23).

FIGURE 2.22

FIGURE 2.23

PROJECT 6 Spin-Art T-Shirts

COST: $$

MAKE TIME: 60 minutes to make the spin-art machine, 5–15 minutes for the shirts, plus drying time

I am somewhat obsessed with spin art. Every summer I pull out my salad spinner and create fun artwork with centripetal force. Physics plus crafting! What could be more fun?

Giant spin art, that's what! Years ago, I came across an Instructables describing how to make a spin-art machine from a box fan (https://www.instructables.com/id/Homemade-Spin-Art-Machine/). Basically, you remove the grills from the fan, trim back the fan blades, and attach a cardboard base. It's surprisingly easy to do.

The giant spin-art machine is a ton of fun for creating artwork. Usually I attach paper plates with tacks and let the campers add paint. Then one year I decided to mount a shirt on the spin-art machine. The results were fantastic. So for Maker Camp that year, we created spin-art Makey T-shirts.

Materials

- 16- to 20-inch-wide box fan
- Cardboard (I used the box the fan was in)
- 4 drywall screws
- Philips head screwdriver
- Electric rotary tool, such as a dremel
- Goggles
- 4- × 4- × 5⁄16-inch carriage bolts
- 8- × 5⁄16-inch washers
- 4- × 5⁄16-inch wing nuts
- Large binder clips
- Masking tape
- Your stencil cut on plastic stencil material, cardstock, or freezer paper
- A cotton T-shirt, washed and dried
- Acrylic fabric paints
- Large box to block paint splatter
- Goggles

Build the Spin-Art Machine

1. Use the screwdriver to remove the grill from the front of the fan.
2. With the rotary tool, cut the fan blades back, leaving as little blade as possible near the center. You don't need to worry about sanding it or making it look nice. You won't see it.
3. Layer two circular pieces of cardboard, roughly the diameter of the fan itself. If desired, glue or tape them together.
4. Use the screws to attach the carboard to the center of the fan. This may take a few tries to get a spot that holds well and does not block the movement of the fan.
5. Using a tape measure, divide the circle into quarters. Use the screwdriver or an awl to punch holes near the outer edge of the carboard on the line between each quadrant.
6. Place a washer on a bolt. Push the bolt up from the bottom of the carboard into a hole. Add a washer and tighten using a wing nut. Repeat with the other bolts.
7. Test the speed of the fan. Make sure that it doesn't wobble. Once you have a setting that you prefer, leave it set. Plug your fan into a power strip to more easily turn it on and off.
8. Place the machine in a large carboard box, and pull up the top pieces to create higher sides, using tape to hold the cardboard in place. Pull the power cord out through the side of the box (Figure 2.24).

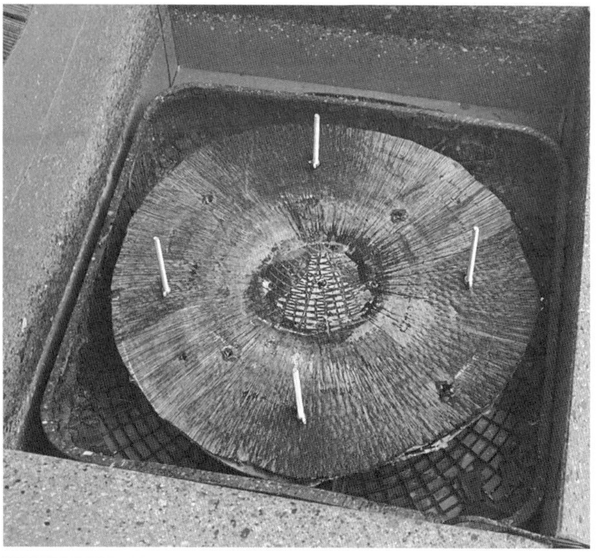

FIGURE 2.24

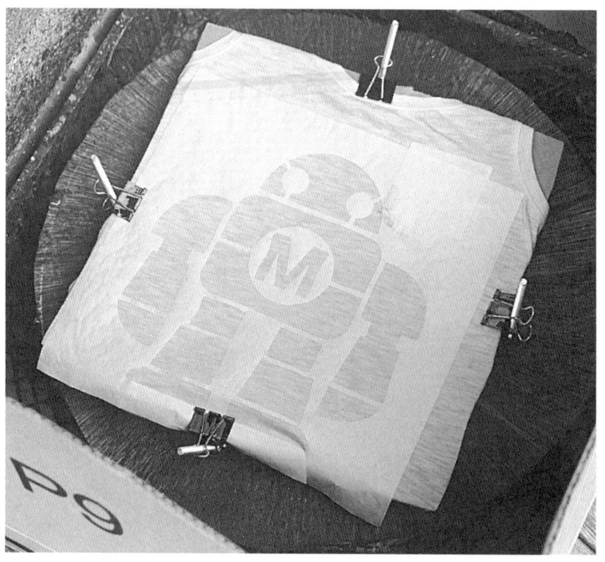

FIGURE 2.25

Paint the Shirt

1. Apply your stencil to the shirt as described previously.
2. Fold back the sleeves, and use masking tape to attach them to the back of the shirt. Fold the bottom of the shirt up onto the back and tape it down as well. If desired, use newspaper or masking tape to cover any area on which you do not want paint splatter.
3. Attach binder clips to the top, bottom, and right and left sides of the cardboard and shirt. Then slide the metal loops on the binder clips over the bolts, and lower the shirt to the cardboard base of the spin-art machine. You may need to adjust the binder clips (Figure 2.25).
4. Put on goggles. Start the machine. From above and aiming toward the center of the shirt, squirt paint onto the shirt. It is best to add a little at a time and to try to cover the entire area. You can turn off the machine as needed to check your progress (Figures 2.26, 2.27, and 2.28).

FIGURE 2.26

FIGURE 2.27

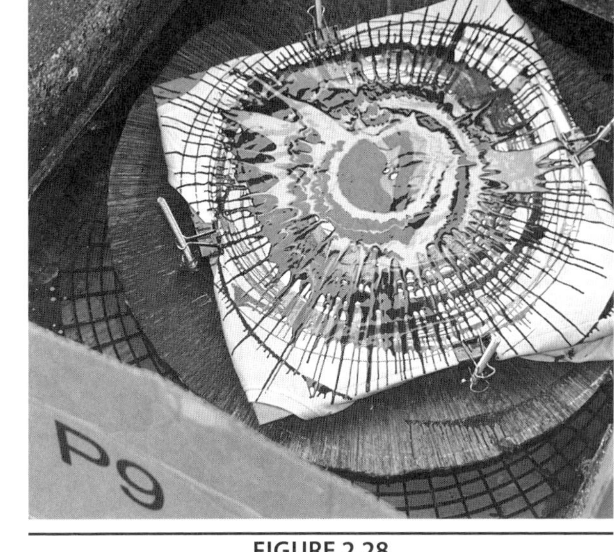

FIGURE 2.28

5. Carefully remove the shirt from the machine. Remove the binder clips. Slowly pull off the stencil (Figure 2.29).
6. Let the shirt dry completely before removing the masking tape from the back and pulling out the carboard.
7. To heat-set the ink onto the shirt, turn it inside out and put it in a dryer for 30–60 minutes. Don't wash the shirt for a week to allow the paint time to cure.

FIGURE 2.29

PROJECT 7
Dirt Shirts

COST: $

MAKE TIME: 5–15 minutes, plus drying time

At World Maker Faire 2017, Loop of The Loom hosted a station that featured Bengala dyeing (https://makerfaire.com/maker/entry/63216/). The process, which comes from India, relies on the natural iron oxide found in soil to help affix color to fabric. The results are beautiful, soft tones unlike anything else. Similar ancient dyeing techniques, such as Bingata and Dorozome from Japan and Bogolan from Africa, use the variations in soil combined with plants, herbs, and spice to create unique fabric colors.

So I wanted to come up with a simple way to explore dyeing our T-shirts with dirt for Maker Camp. To give campers the ability to customize with dirt shirts, offer richly colored spices such as turmeric, paprika, curry powder, mustard, cinnamon, and the like to mix in with the dirt. These are easy to get in bulk and tend to be inexpensive. You can also try adding ground flowers, dried herbs, onion skins, coffee, tea, or other natural materials to your mix.

If your soil has no red or orange hue, it may be low in iron oxide, but that's easy to correct. You can buy iron oxide from most scientific supply companies and add it when you add the spices.

There are a lot of ideas on the best ways to fix the dye to fabric. A rinse in soda ash (sodium carbonate), as you would do for tie dyeing with synthetic dyes, is most common. You can buy soda ash online. Alternatively, you can pick up Arm & Hammer Washing Soda in many grocery stores or pH Up from a swimming pool supply store. Soda ash prepared for dyeing applications contains less water and is more concentrated, so plan to use twice as much of the powder when making your prefix solution. You can skip the prefix step below, but this will result in less vibrant colors that may fade more quickly.

When mixing your dirt paint, you can simply use warm water. Sometimes boiled soybean liquid is used to help transfer the dyes, and I tried using store-bought soymilk as a liquid. I found that the soymilk created less bleed but didn't significantly brighten the colors. It may be worth using if you choose not to use the prefix step. You can also use vinegar as your liquid. Vinegar will adjust the pH and help your dye attach to the fabric without fading.

Materials

- Your stencil cut from plastic stencil material, cardstock, or freezer paper
- A cotton T-shirt, washed and dried
- A trowel or shovel
- A sieve or colander
- Plastic containers for mixing dirt paint
- Spoons
- Assorted spices such as turmeric and paprika
- Soda ash
- Water
- Disposable gloves
- Foam paint brushes

Applying Prefix

1. Put on disposable gloves. Add ½ cup of soda ash or washing soda to 1 gallon of warm water. Mix well.
2. Place your shirt in the prefix solution. Let it soak for at least 5 minutes and up to 1 hour.
3. Wring out the shirt. Either line dry it or dry it in a dryer. Do not rinse the shirt.

Mixing Dirt Paint

1. Use your trowel or shovel to collect soil. Remove any plant materials. Pass the soil through the sieve or colander to remove clumps or rocks.
2. Put about 1 cup of dirt into a plastic container or bowl. Add iron oxide and spices, if you are using them (Figure 2.30).

FIGURE 2.30

3. Add water or vinegar a little at a time, mixing with each addition. Continue until you have a thick paint consistency. Add more dirt if the paint gets too thin (Figure 2.31).

FIGURE 2.31

Applying the Dirt Paint

1. Attach your stencil to your shirt.
2. Using your foam brush, apply the dirt paint over the stencil (Figure 2.32).

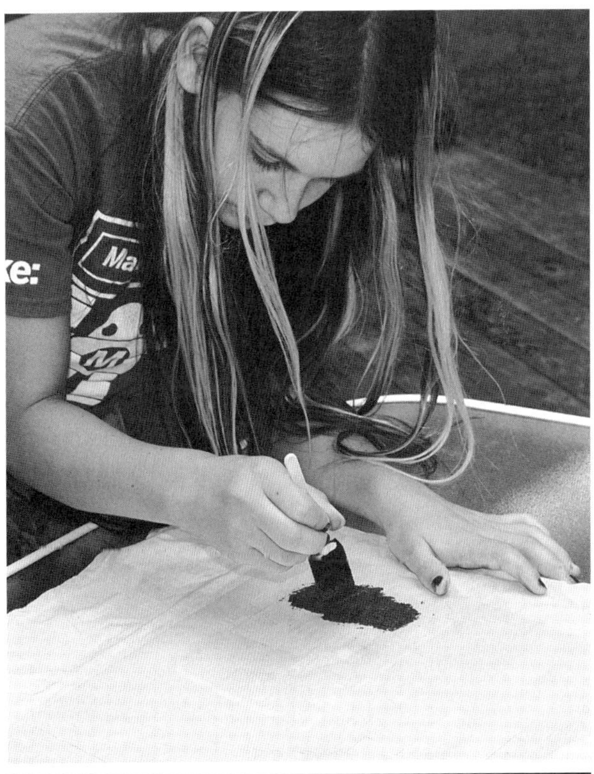

FIGURE 2.32

FIGURE 2.33

3. Carefully remove the stencil (Figure 2.33).
4. Let the dirt paint dry on the shirt for at least 60 minutes (overnight is even better; Figure 2.34).

FIGURE 2.34

5. Shake off as much of the dirt paint as you can, and then rinse the shirt in cool water until the water runs clear. Dry the shirt well. To heat-set your design, turn the shirt inside out and put it in a dryer for 30–60 minutes (Figure 2.35).
6. Add additional decorations as desired.

FIGURE 2.35

CHAPTER 3

Camp Wearables

Although I love T-shirts, there are other great wearables that kids love to make. Some are definitely science centered, with great opportunities to learn about light, electricity, and more. Others are more whimsical, like creating costume play (*cosplay*) characters. Either way, these creative projects will give your campers many opportunities to express who they are.

PROJECT 1
Ultraviolet (UV) Color-Changing Sun Bracelet

COST: $

MAKE TIME: 5–10 minutes

We all know that protecting our skin from the sun is important for our health, but kids are often too busy having fun outdoors to think about reapplying sunscreen. These bracelets provide a simple—and fun—reminder to take care while getting some rays.

This project relies on beads with UV-sensitive pigments. The colorless dye molecule physically changes when exposed to UV light. When exposed to sunlight or light from another UV source, the white beads change to many different colors. When the UV light is removed, the colorful beads return to white. This cycle can be repeated hundreds of times (Figures 3.1 and 3.2).

Before making the bracelets, why not take some time out for science? Have your campers test the beads to determine which materials protect the beads from dangerous UV and which don't. For example, place the beads under water, behind sunglasses, or in the shade. Cover them in different sunscreens. Take note of what stops, diminishes, or slows down the color change.

The Stanford SOLAR System, an educational outreach organization funded by NASA as part of the Solar Observatories Group, has great lesson plans and resources available if you'd like to spend some time working through the science (see http://solar-center.stanford.edu/activities/UVBeads/UV-Bead-Instructions.pdf).

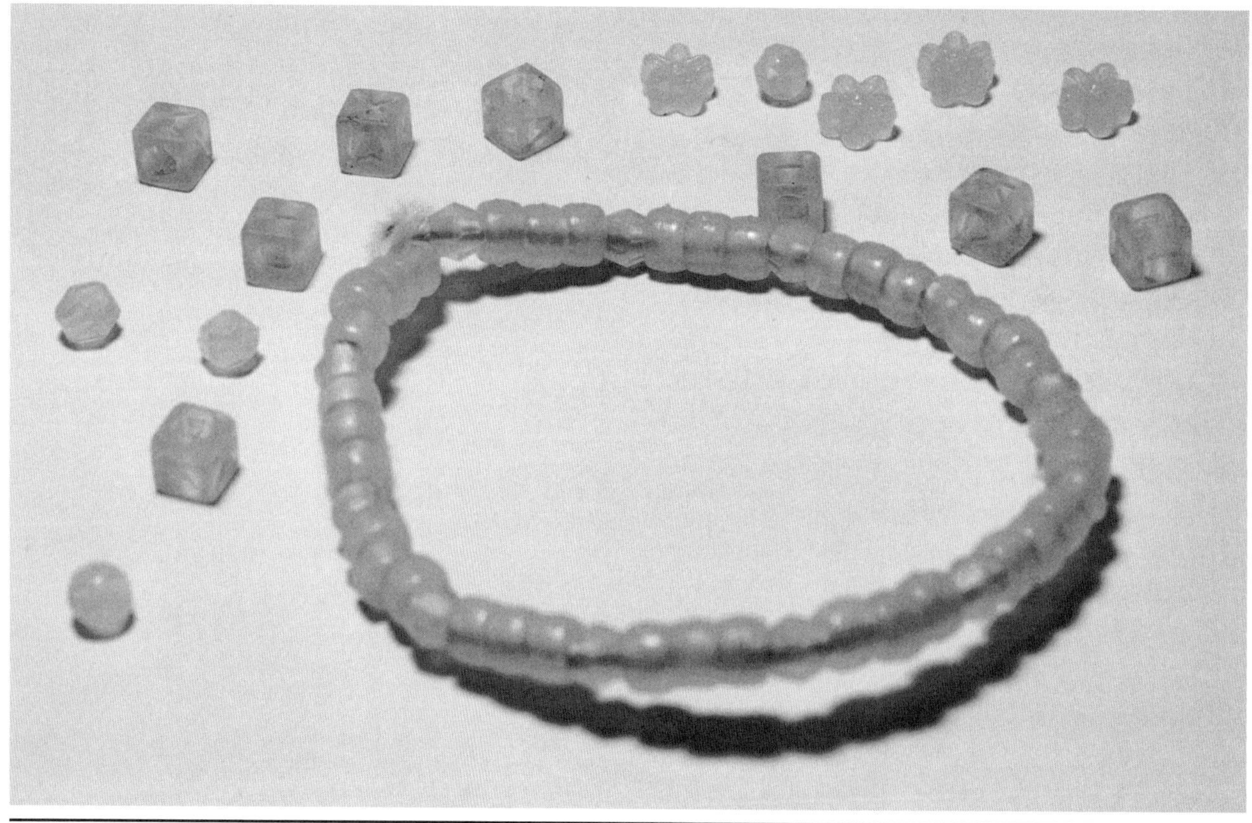

FIGURE 3.1

FIGURE 3.2

Materials

- Ultraviolet-detecting beads, such as those offered by Educational Innovations, Inc.
- Chenille stems (pipe cleaners)
- Scissors or wire cutters
- A portable long-wave UV light source or flashlight (optional)

Process

1. Wrap a chenille stem loosely around the wrist of each camper to judge the size needed. Use the scissors or wire cutters to trim the stem as needed.
2. Use sunlight or a UV flashlight to see the colors of the beads. String the beads on the chenille stem. Twist the ends of the stem together to finish the bracelet.

PROJECT 2 Simple LED Crafts

COST: $

MAKE TIME: 5–10 minutes

Light-emitting diode (LED) "throwies" are a popular makerspace project. The concept is simple: You combine a small light with a coin cell battery and attach a magnet. Originally, mischievous makers would throw them at buildings so that they attached to the gutters. That's probably not something you want to encourage your campers to do, but you can use the same concept to make other fun items.

Obviously, light-up magnets for the fridge would be fun, but so are wearables such as pins, barrettes, and clips. With a strip of felt and a few snaps, you can create a glowing bracelet. Or you can use larger LEDs (10 millimeters) to bring a bit of magic to camp and create magic wands by twisting together a few chenille stems and placing the LED on top (Figure 3.3).

FIGURE 3.3

Materials

- 5-millimeter LEDs in assorted colors
- 3-volt lithium coin cell battery (CR2032 or CR2025)
- Colorful duct tape
- Hot glue or craft glue
- Assorted craft supplies such as feathers, chenille stems, plastic jewels, sequins, felt shapes, googly eyes, ribbon, etc.
- Pin-back clasp brooches, French barrette hair clips, and/or alligator hair clips

Process

1. The long leg of the LED is the positive lead. The short leg is the negative lead. Slip the LED over the battery so that the positive lead is over the smooth, positive side of the battery and the negative lead is over the rough, negative side of the battery. The LED should light up.
2. Use duct tape to attach the battery and LED.
3. Decorate with craft supplies as desired.
4. Use hot glue, craft glue, or duct tape to attach the decorations (Figures 3.4 and 3.5).

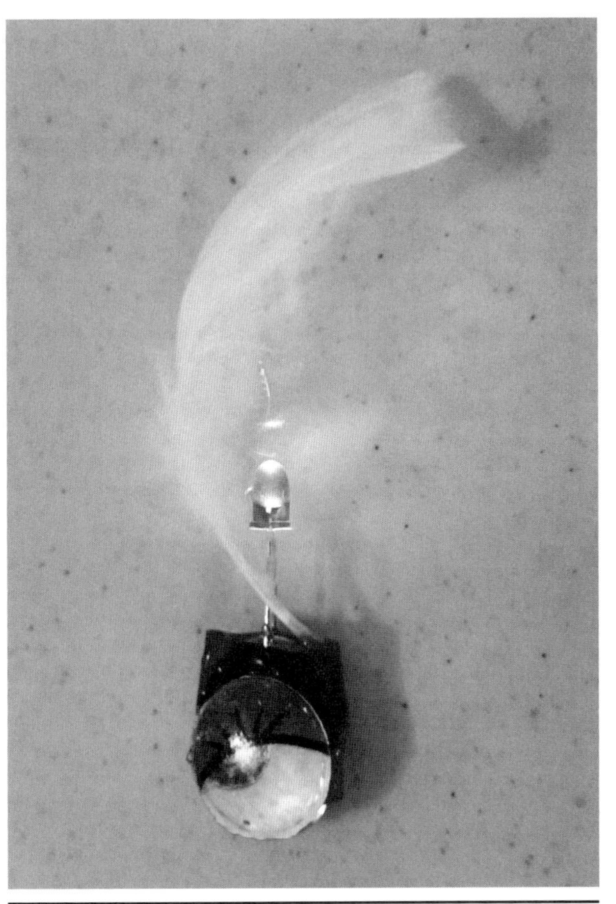

FIGURE 3.4

FIGURE 3.5

PROJECT 3
Light-up Proximity Friendship Necklace

COST: $$$

MAKE TIME: 30–60 minutes

Friendship bracelets are a classic camp craft. Much like the ubiquitous lanyard or macramé plant hanger, traditional friendship bracelets are usually made by tying knots with embroidery thread. The colorful completed bracelet is given to a friend. I still have a collection from my camp days.

But I think it may be time for an update. Let's make our friendship-ware interactive! In this project, we'll use the infrared radiation (IR) capabilities of the Circuit Playground Express (CPX) to make a device that will recognize a friend when they come near and alert you to their presence with flashing lights and sounds. The CPX was designed by Limor "Ladyada" Fried for Adafruit Industries. The goal was to create a microcontroller that was easy to use, easy to program, and affordable. It comes loaded with sensors, programmable lights, a speaker for music, and more. Programming is easy using the free, open source, visual drag-and-drop Microsoft MakeCode online platform as well as popular languages such as JavaScript and Circuit Python. To learn more about the CPX and explore additional projects, check out *Getting Started with Adafruit Circuit Playground Express* by Mike Barela (Maker Media, Inc., 2018).

Infrared radiation (IR) is a common light source used in technology such as TV remote controls. The wavelength of IR light is longer than that of visible light, so IR is invisible to the human eye. However, it can be used for great line-of-sight wireless communication, sending short amounts of data between devices. The CPX has a range of 10 to 20 meters. This project is designed for two besties in close range but can be easily expanded to include additional CPX boards if you're more of a social butterfly.

Once your CPXs are coded, you'll need to find a way to wear them. One of the easiest is a magnetic clip or single pin back attached right onto your awesome Maker Camp T-shirt. Of course, using a few double-sided adhesive foam squares and craft supplies, you can easily make a pretty neat necklace with the CPX. If you have access to a three-dimensional (3D) printer, check out the Thingiverse (https://www.thingiverse.com/) for CPX holders you can modify and print.

This project's code was originally based on an interactive Zombie Tag game by Kattni Rembor and Carter Nelson of Adafruit Industries. This game may also be a hit with your campers, so be sure to check it out (https://learn.adafruit.com/circuit-playground-express-ir-zombie-game/makecode-zombies). You will need at least three CPX devices for the Zombie Tag game.

We'll be coding in MakeCode, a visual language that uses interlocking blocks of code to create computer programs. Each group of blocks is color-coded by purpose. The window is divided into three sections: On the left there is an interactive model of the CPX board that will virtually follow your code, in the center is a panel with the code blacks, and on the right is the script area, where you will arrange your code (Figure 3.6).

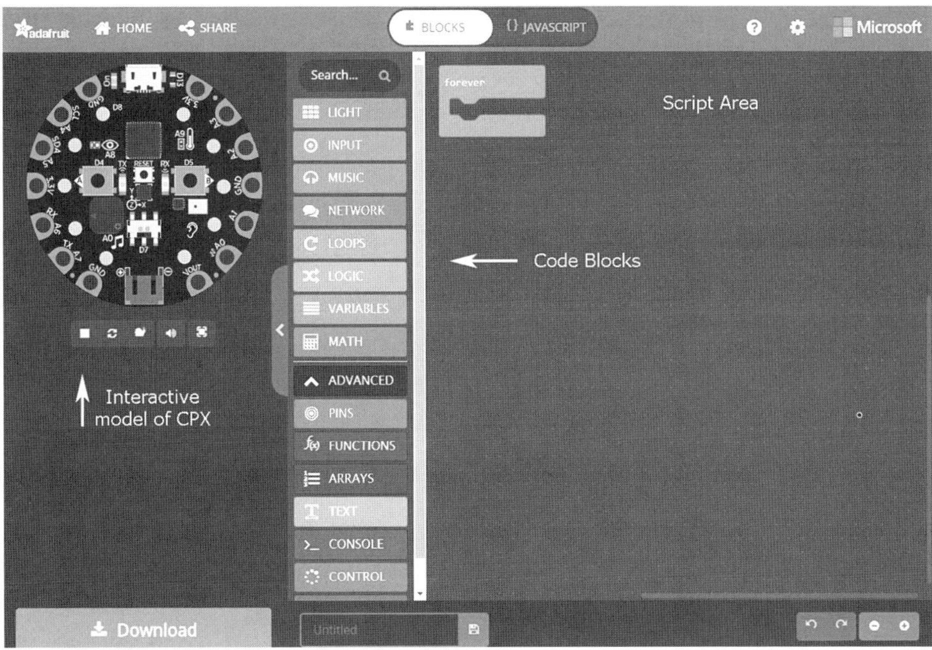

FIGURE 3.6

Materials

- At least two Circuit Playground Express (CPX) boards
- Three AAA battery holders with on/off switch and JST 2-pin connector cable
- Mini USB cables

Process

1. Go to https://makecode.adafruit.com/. Click on "New Project."
2. Click on the green "Variables" menu. Click on "Make a variable." Create the following variables: "Me," "Friend1," "Friend2," "Whoisit," "Switch," "Timeon," and "Time2next."
3. Click on the green "Loops" menu. Drag the "on start" block to your script area. This loop will initialize, or set up, your CPX.
    a. For each of these variables, go to the "Variables" menu and drag the pink "set to" block into the script area. Place the block into the "on start" block. There is a little downward-facing arrow. Clicking on it will show a drop-down menu of all your variables. Use this to set the variable name. The white oval has a number in it. Click on the white oval to change it.
    b. The "Me" variable represents the person wearing this CPX. Set this value to 0.
    c. The variables "Me," "Friend1," and "Friend2" represent each of our friends. I suggest marking each CPX with a number. The number of the CPX you are wearing will be your number in the list of variables. Set "Me" to the number you are wearing. Set the other variables to the appropriate numbers. You can use your friends' names rather than the generic "Friends" variables in the example code.
    d. Set "Switch" to 1. This will control whether the CPX is receiving and sending signals at all. Basically, it will help us code an on/off switch for IR.

e. Set "Brightness" to a level between 1 and 255. I set it to 10 because I want to conserve power, but you can make your necklace as bright as you like.

f. Set "Time2next" to 0. We will use this variable to control how often the CPX alerts you to your friend's arrival.

g. From the blue "Light" menu, select the "set all pixels to" block, and place it inside the "on start" block. Select a color. All the lights on the CPX will glow this color.

4. Let's tell the CPX what to do when it receives a signal from a friend.

   a. From the "Network" menu, select the "on infrared received" block, and drag it onto the script area. Use the drop-down menu to select the variable "Whoisit."

   b. From the "Variables" menu, drag the "set to" block into the script area, and place it into the "if then" block. Use the drop-down menu to select the variable "Timeon."

   c. Under the shown menus is a button that reads "Advanced." Click it. From the dark teal "Control" menu, select the "millis (ms)" oval, and drag it into the "set to" variable block, placing it into the white oval. This block tracks how long it has been since the CPX was turned on. We'll use this as a timer to control how often the CPX responds after finding your friend. Otherwise, it will go off over and over again very quickly, which can get annoying.

4. Let's code the on/switches next. The CPX has two buttons on the front. The one on the left is labeled "A." The one on the right is labeled "B."

   a. From the purple "Input" menu, drag the "on button click" block into the script area. From the drop-down menu, select "Button A." Leave the second on "click." From the "Variables" menu, drag the "set to" block into the script area and place it inside the "on button click" block. Set the drop-down menu to "Switch" and the number to "1." This will turn the IR on.

   b. Repeat the preceding, but set the "on button click" to "Button B" and the number for "Switch" to "0." This will turn the IR off.

5. Okay. Now it is time to code the main functions of the CPX. First, we're going to write code to tell the CPX to send out a signal so that your best friend can find you.

   a. From the green "Loops" menu, select "forever," and drag it onto the script area.

   b. From the same menu, select the "while" loop, and place it into the "forever" loop.

   c. From the "Logic" menu, drag the "=" block in, and place it into the "true" area of the "while" block. Place the variable "output" into the first white oval, and then enter "1" into the second white oval. This means that anything inside the "while" block will happen as long as the "Switch" variable is set to 1 (in other words, when the IR is turned on). If "Switch" is set to 0, or any other number, the loop will not send out a signal.

   d. From the "Network" menu, select the "infrared send number" block, and put it into the "while" block. Set the drop-down menu to "Friend1." This will send your number out to other CPXs.

e. From the "Loop" menu, drag the "pause" block into the "while" loop. Click on the white oval's drop-down menu, and select "1 second." You can always adjust this later, but putting a pause in here conserves power by slowing down the signal your friend can react to.

6. So what will actually happen when your CPX finds a friend? Your CPX will read the signal and react with color and sound, but the CPX works very quickly, so we have to slow it down a bit first so that we don't get blasted with noise and light. Let's code it.

   a. From the "Logic" menu, select the "if then" block, and place it inside the "forever" loop and under the "while" loop that we coded earlier.

   b. Select the "<" block, and replace "true" in the "if then" block. Set the first oval to the variable "Time2next" and the second to "Timeon." This will check to see if "Time2next" is less than "Timeon." Because we initially set "Time2next" to 0 and "Timeon" to the number of milliseconds the CPX has been powered, that statement will be true, and the rest of the loop will run.

   c. From the "Variables" menu, select "set to," and place it inside the "if then" block. Set the drop-down menu to "time2next." From the dark purple "Math" menu, select the "+" oval. Drag it into the white oval of the "set to" block. In the first white oval, place the variable "Timeon." In the second oval, type "60000." This line of code will add 60 seconds to the "Time2next" variable. So rather than having music blaring and the lights flashing constantly, they will perform their animation and then pause for a minute, giving you time to hit the "off" switch while chatting with your friend.

7. Now it is time to code the response to your friend. We're going to play a sound and then flash an animation that is different for each friend. You can customize both of these options.

   a. From the light blue "Logic" menu, select the "if then" block, and place it under the "set to" variable block, inside the first "if then" loop. This will create what we call a *nested loop*. The code will check the first condition of the first "if then" loop. If that condition isn't met, it moves to the next loop.

   b. From the "Logic" menu, drag the "=" triangular block into the script area, and place it into the space presently marked "true." The block will replace "true."

   c. From the "Variables" menu, drag the oval for "Whoisit" into the first white oval on the "=" block. Drag "Friend1" into the second white oval. The block should read "if whoisit = Friend1 then." This will be the animation for your "Friend1."

   d. From the "Loops" menu, drag the "repeat" block into the "if then" block. Set the repeat to five times. This will mean that your animation will run for about 5 seconds. You can make it more or less by adjusting the number of repeats.

   e. From the "Music" menu, drag the "play tone at" block into the "repeat" loop. In the first oval, type "10000." In the drop-down menu, select "¼ beat." We've set the frequency of this sound very high, higher than the notes and preprogrammed sounds that come with the CPX. This is because the IR interferes with the sound on the CPX, so we need to push the sound above that sign noise to get a nice-sounding note. By the way, if you want to use several tones to make a melody here, you can.

f. Add the "stop all sounds" block after your selected music.

g. From the "Light" menu, add the "set brightness" block to the "repeat" loop, and change the number to "255," the maximum brightness.

h. From the "Light" menu, add the "show animation" block. Select an animation for this friend. On the drop-down menu, set the time to "1 second." Here again, you can design your own animation if you prefer.

i. To add the next friend, repeat steps a–h. Change the variable that matches "Whosit" to "Friend 2." Change the tone you make and the animation.

j. After the last friend "if then" loop, but inside the "time" loop, add the "Light" block "set all pixels to" and "set brightness." Reset the CPS to your initial color and brightness selections (Figure 3.7).

7. Each CPX will need to have its customized code loaded onto the device.

    a. Give your code a name, such as "Friend bracelet 1." Click the "Save" icon (the small floppy disk), and then click "Download." A .uf2 file will be downloaded to your computer.

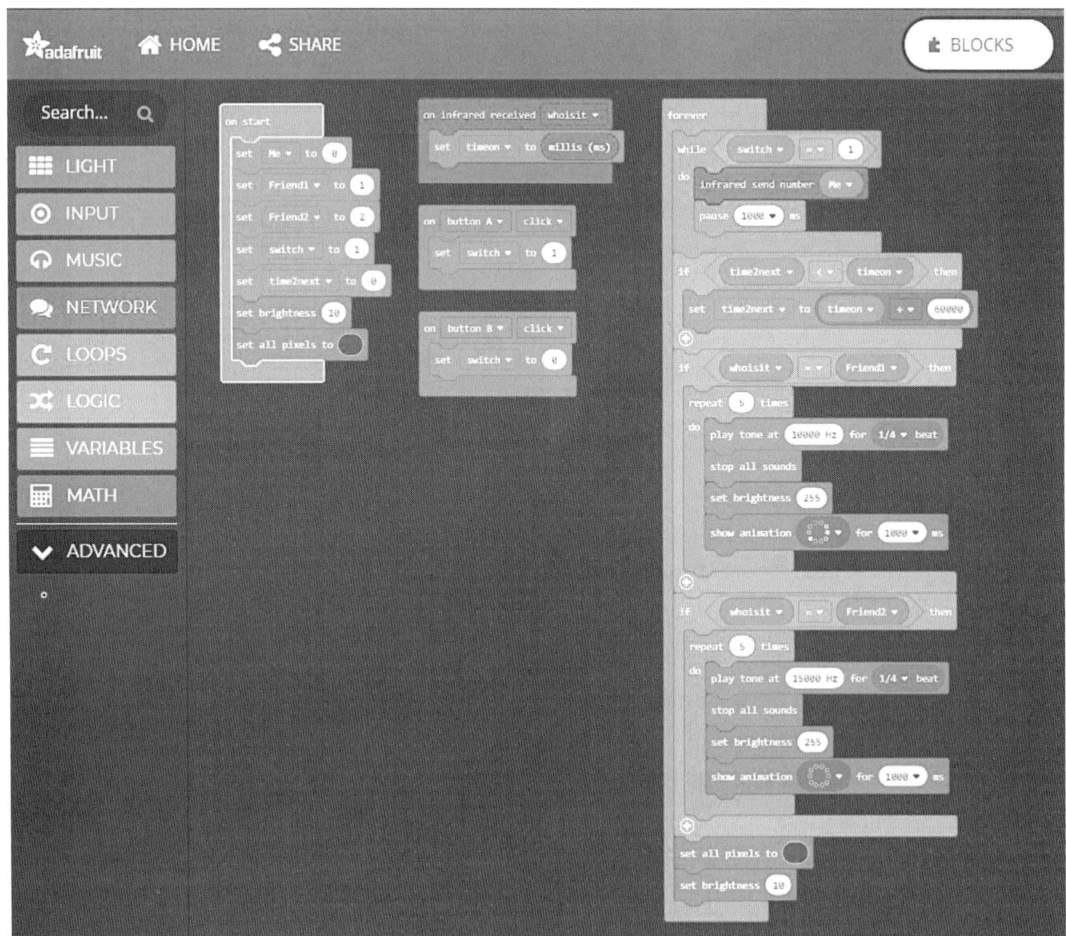

FIGURE 3.7

b. Use the USB cable to plug the CPX into your computer. Tap the small "Reset" button in the center of the CPX.

c. Look for the CPLAYBOOT drive in File Explorer or Finder. Copy the .uf2 file from your computer onto the CPLAYBOOT drive by dragging it over.

8. If desired, use an adhesive-backed foam square to attach your CPX to the necklace you have crafted. Use another foam square to attach a battery pack to the back of the necklace. Turn on your CPX, and test it out (Figure 3.8).

NOTE: The complete code is available for download at https://makecode.com/_P5MVgDLxd9W7.

Take It Further

If you are planning to have a lot of friends with CPXs, or if you are planning complex music or animations, you may want to use "Functions" to help organize your code. Here's what you need to do:

1. Click on the "Advanced" button. Then click on the royal blue "Functions" menu.
2. Click on "Make a function," and create a function for each friend. A "function" block will automatically appear in the script area.
3. Select on a friend's code. Drag the "repeat" loop and everything in it into the "function" block. Leave the "if then" for that friend's variable in the large "forever" loop.
4. Repeat step 3 for each friend.
5. From the "Forever" menu, select the "call function" block and place it within the "if then" loop for each friend.
6. Match the variable called by the "call function" block to the variable in the "if then" loop.

   See the sample code for help.

You can also program a simple animation for your own CPX badge that runs all the time. Simply use a "forever" loop and blocks from the "Light" menu to code it.

FIGURE 3.8

PROJECT 4
Stamped Leather Bracelet

COST: $$

MAKE TIME: 5–10 minutes

Maker Camp is a great opportunity to practice traditional hand crafts. Leather working is a fun way to create exciting wearables while using tools and working with materials many students don't often have the opportunity to use. This leather bracelet is always a hit. Not only do the campers love to use hammers and make some noise, they love the results.

You will want to do a basic talk about hammer safety before beginning this activity. Roll up long sleeves and tuck in shirts. Tie back long hair for maximum visibility and safety. Remind campers not to run with tools and to treat tools with respect. If you are using hammers, keep them clean, and inspect the handles before beginning. You may even want to put a tennis ball over the claw.

Campers will need to learn to swing the hammer correctly. Remind them to always check around them and behind them before starting so that no one is accidently hit. Show campers how to grip the hammer tightly at the end of the handle, with their thumb placed for stabilization. They should line up the hammer and item before swinging for an accurate strike. Campers should swing the hammer with their entire arm using moderate, consistent force. You may want them to practice first.

Materials

- Blank leather wristbands, like those manufactured by Realeather Crafts and sold in craft stores
- Metal leather stamping sets, at least one alphabet plus additional shapes

NOTE: If possible, choose sets that have built-in handles rather than a set that has several stamps and just one snap-in handle. This allows more campers to work at once.

- Rubber mallets or lightweight hammers
- Blocks or planks of wood
- Safety goggles
- Leather awls
- Dye ink pads

Process

1. Have campers take a moment to plan their design. This is especially important if they plan to add their name or other words.
2. Place the wristband on the wood block or plank. It can help some campers to use rubber bands or tape to hold the wristband in place. Dampen the leather with a little water to make it more pliable.
3. Place a stamp against the leather so that the stamp is perpendicular to the leather surface. Hold the stamp close to the leather surface.
4. With the mallet or hammer, whack the top of the stamp. It is better to use a few strong swings rather than many taps because the stamp can shift.
5. Once the stamp design is done, use the awl to add punched holes or to scratch in additional designs, if desired.
6. Add color to your creation by gently rubbing the wristband over the ink pads, if desired (Figure 3.9).

FIGURE 3.9

Take It Further

Try designing and printing your own leather stamps using a computer-aided design (CAD) program and a 3D printer. You'll want to design your print to be about $1/8$ inch deep (approximately 3 millimeters) with a $1/8$-inch backer plate. Print in ABS for additional strength. Clear line art designs are best.

For a complete discussion on using SketchUp to create 3D printed leather stamps, see Matt Makes Instructables on the topic at https://www.instructables.com/id/How-to-Design-Print-and-Use-3D-Printed-Leather-Sta/.

PROJECT 5
Light-up Sunglasses

COST: $

MAKE TIME: 15 minutes

If it's summertime, you need to have sunglasses. If you're going to have sunglasses, they ought to be really cool. If you want them to be really cool, make them light up. It's a pretty simple plan.

You can pick up inexpensive sunglasses at your local dollar store. I usually try to get plain black glasses so that they are a neutral canvas for campers. Look for glasses that have thick rims and wide bridges and arms.

Glow-in-the-Dark Painted Sunglasses

Did you know that they make a glow-in-the-dark neon-colored dimensional paint? They do! So channel your inner child of the '80s and make some "totally rad" shades. Then break out the black light, spin some Cyndi Lauper, and call it a party!

Materials

- Inexpensive sunglasses
- Neon glow-in-the-dark dimensional paint in fine-tipped squeeze bottles
- Masking tape
- Dish soap
- Blow dryer (optional)

Process

1. Wash the sunglasses with dish soap and water. Dry well.
2. Cover the lenses with masking tape or remove the lenses.
3. Use the paint to add dots, swirls, and stripes to the rims, bridge, and arms of the glasses.
4. Allow the paint to dry or use a blow dryer set to cool to speed up drying (Figure 3.10).

FIGURE 3.10

Glow Stick Sunglasses Hack

This is a really easy and inexpensive way to make glowing sunglasses. Use them for evening events such as the Fourth of July fireworks or stargazing nights. These are also really fun for long-exposure photography with a group of campers in a darkened room. Use the same technique to add glowing elements to gloves, hats, ties, or bows, and then turn out the lights and have campers pantomime a story while capturing it on camera. This can make being part of a video project more comfortable for shy performers.

Materials

- Inexpensive sunglasses
- Thin glow sticks, the kind used for bracelets
- Cable ties/zip ties
- Diagonal cutters or scissors
- Sandpaper or nail file

Process

1. Crack the glow stick to make it glow. Make sure that the entire stick is glowing well. Shake it to ensure that the fluid is well distributed.
2. Use your hands to gently warm the glow stick, bending it along its entire length to make sure that it is as flexible as possible.
3. Wrap a zip tie around the bridge of the sunglasses. Insert the tapered, ridged end into the receiving end, and tighten slightly to create a loop.
4. Place the glow stick along the top rim of the sunglasses, sliding it into the loop. Center it over the glasses. Tighten the zip tie, making sure that the receiving end is to the top and back of the glasses. Leave a little wiggle room so that you can make adjustments as needed.
5. Use zip ties to attach the glow stick at the end of the rim near the arm (next to the hinge) on each side. Make sure that the receiving end is to the top and back of the glasses. Leave a little wiggle room so that you can make adjustments as needed.
6. It is likely the glow stick will be longer than the width of the rim of your glasses. As we did earlier, wrap a zip tie around the arm of the sunglasses. Insert the tapered, ridged end into the receiving end, and tighten slightly to create a loop. Gently bend the glow stick, and tighten the zip tie to hold in place along the top of the arm.
7. Adjust the glow stick and zip ties as needed, and tighten them all to secure the glow stick.
8. Use the diagonal cutter to snip the excess zip tie material (the tapered end). If needed, use the sandpaper or nail file to remove excess material so that the glasses are comfortable to wear (Figure 3.11).

FIGURE 3.11

Light-up Sunglasses

COST: $

MAKE TIME: 30–45 minutes

A long time ago, in my box of disguises for Maker Camps, I had a pair of neon green sunglasses that featured a blinking LED in the center. They were, obviously, a favorite among the campers, which means that the battery, which could not be replaced, died quickly. The glasses were soon broken and discarded. It was sad.

So I decided that I would make an even better pair of light-up sunglasses with more lights and a battery that could be replaced. Guess what? It wasn't easy to do! Believe it or not, this single project went through more prototyping than any other project in this book. In the end, I prevailed, and now you too can make awesome light-up sunglasses. My point is: follow the directions. Trust me.

Regarding the LEDs themselves, you have to plan carefully if you want to use more than one. This is due to the voltage needed to power the LEDs. You can use multiple LEDs of the same color without an issue. You can also pair red and yellow or blue and green LEDs. But red and blue together will not work unless you wire in a resistor because the red uses less voltage than the blue, so the circuit will bypass the blue and run only the red, taking the path of least resistance. (See the "Light-up Letter Home" in Chapter 4 for a detailed discussion on circuits.)

Test, test, test your circuit with alligator-clip leads before you apply it to the sunglasses to ensure that your parallel circuit will power all your LEDs with 3 volts. This will save you a lot of headaches in the long run. Again, trust me on this.

As a side note, LEDs are available that blink in single colors and that blink red, green, and blue. If your budget allows for it, you may want to buy the multicolored LEDs for this project. They aren't that much more expensive than traditional LEDs. You may also want to consider using LEDs designed for wearable electronic platforms such as LilyPad (https://www.sparkfun.com/about_lilypad) or Gemma (https://learn.adafruit.com/introducing-gemma). These will be the easiest to use because you can glue them directly to your sunglasses, and they are sturdy so there is no chance of breaking a lead, which is a major concern in projects like this. However, they are roughly three times the cost of traditional LEDs.

In this project, we use a battery holder meant for a sewn wearable project. It sits flat against the arm of the sunglasses and provides a lead on each end, to which you can attach your wires. The battery is slipped in and holds well. The only downside is that there is no switch.

As for the wire, we will be using a thin wire meant for beading. It's inexpensive, easy to find at any craft store, and comes in many colors. Don't use anything over 28 gauge. It won't be flexible enough for this project. When selecting your wire, make sure that it is not nylon or paint coated or a beading string that is mixed with nylon fiber. You want 100 percent metal. This type of wire is also sold as "hook up" wire or magnet wire (because it can also be used to make electromagnets).

Materials

- Inexpensive sunglasses
- 3-millimeter LEDs
- 3-volt coin batteries (Product Number 2032 or 2025)
- Sewable CR2032 battery holder
- Metal wire, 28 gauge
- Drill or rotary tool with ⅛-inch drill bit
- Sandpaper or nail file
- Needle-nose pliers
- Wire cutters
- Super glue
- Hot glue
- Small binder clips
- Metallic permanent marker, paint pen, or white-out, for marking LED locations
- Safety goggles
- Scrap wood
- Masking tape
- Alligator-clip leads
- Double-sided adhesive foam squares

Process

1. Determine where you would like to place your LEDs. One in the center of the bridge is a good location. You can also add LEDs to the upper part of the rim, where it gets wider by the arms, on either side of the lenses. On the inside of the glasses, mark where you plan to put the LEDs. Check that the locations are centered and level.
2. Cover your lenses with masking tape or remove them.
3. Place your sunglasses on a piece of scrap wood. Using your drill or rotary tool set to low, drill holes into the sunglasses where you have made marks. Take your time. You

do not want to crack your glasses or deform the soft plastic.

4. Use sandpaper or a nail file to clean your drilled holes, removing any excess material and ensuring a smooth surface for your LEDs. Take care not to damage the plastic of the sunglasses, especially on the front.
5. Wipe down your glasses with a damp rag and dry well.
6. Place the bulb of a LED into each hole from the back of the glasses so that the light will shine in front. Arrange the LEDs so that the longer positive lead is aligned with the top of the glasses and the shorter negative lead is aligned with the bottom. You may want to use a marker to color the positive leads so that they don't get mixed up. *Do not glue your LEDs into place yet.*
7. Using the alligator-clip leads, test each LED and your circuit. Place your battery into the holder. Clip a lead to the positive lead of the holder. Clip the other end of the lead to the positive lead of the nearest LED. Then clip another lead to the same LED lead, attaching the other end to the next LED positive lead and so on, creating a parallel circuit. Repeat for the negative leads. All the LEDs should light up. If they don't, check for shorts or breaks. Check that your LED leads are aligned properly. Make sure that all the LEDs use approximately the same voltage by referring to the documentation that came with the LEDs.
8. Cut a length of wire approximately 1 foot long. On the positive lead of the battery holder, thread the wire through the hole, and twist it to attach the wire to the battery holder. Use needle-nose pliers to gently crimp the wire to the lead to ensure a good connection. Repeat for the negative lead of the battery holder.
9. Use a foam square to attach the battery holder to the arm of the sunglasses horizontally, with the positive lead facing the front of the glasses.

NOTE: We will replace the foam with glue later, but this allows the campers to more easily move the holder if needed.

10. Gently bend the positive lead of the first LED toward the center of the glasses. Take care to not break the lead.
11. Take the positive wire from the battery holder and wrap it around the arm of the glasses once or twice, and then pull it toward the first LED. Gently but tightly wrap the wire around the lead of the LED. Use the needle-nose pliers to gently crimp the wire to the lead.
12. Continue bending the positive leads of each LED against the glasses and wrapping the *same* wire around them, crimping each gently. Make sure that the wire follows the frame of the glasses. Use small binder clips to hold the wire in place as you work. Leave the excess wire for now.
13. Repeat steps 10–12 but use the negative wire. Make sure that at no point do the positive and negative wires or LED leads cross. This will short your circuit, and it won't work.
14. Place the battery into the holder, taking care to note the positive and negative sides. Your LEDs should all light up. If they don't, check to make sure that all the wires are wrapped tightly around the leads of the LEDs and that there are no points at which the positive and negative wires overlap.
15. If your circuit works and everything lights up as it should, it is time to make the circuit permanent. Gently pull out each LED, one at a time, just enough to add a drop of

super glue. This will hold the LED in place in the hole. Wait for all the glue to dry.

16. Check that your circuit still works, and adjust it if needed. Use hot glue to attach the wire to the frame of the glasses. You do not need to cover all the wire, but you can if you prefer. Also reinforce and seal the areas where the LED leads are wrapped. Check each connection to make sure that it works before the glue dries. (I left the battery in the holder while I worked.)
17. Hot glue the battery holder to the arm of the sunglasses.
18. Trim the excess wire.
19. If desired, add a little hot glue to the front of the glasses in front of the LEDs to diffuse the light (Figures 3.12 through 3.15).

FIGURE 3.12

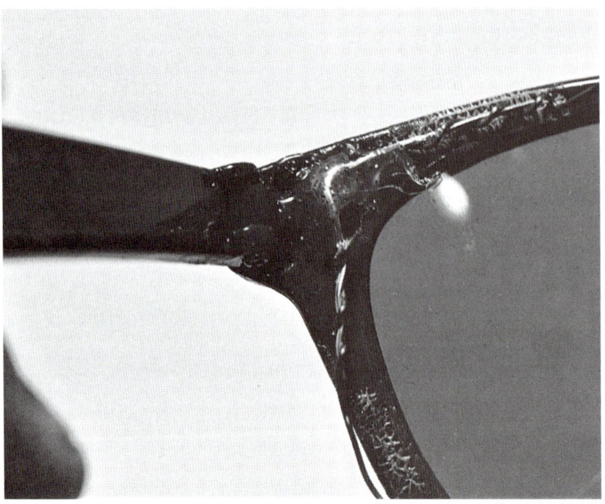

FIGURE 3.13

FIGURE 3.14

FIGURE 3.15

Take It Further

Consider using sewable circuit components for your sunglasses. In addition to the LEDs, sensors and switches are available that you can add to your project. Or try coding using the LilyPad or Gemma. These tiny boards can control the LEDs and create many different lighting effects, including slow and fast blinking, a "twinkling" effect, reactions to sound and light, and movement and more. This really takes your project up a notch. Look into neopixels or other programmable RGB LEDs. These amazing little lights can be coded to change to any color of the rainbow.

PROJECT 6
Cosplay at Camp

In recent years, cosplay—the practice of dressing up as a character from a movie, book, or video game—has gone mainstream. It's like Halloween all year! But seriously, making and wearing a costume that embodies a favorite character—or even better, a character that a camper has created—can be tremendously creative and empowering.

You can explore cosplay in simple ways—paper masks, cardboard accessories, and T-shirt capes—as shown in this section. Or you can take your camp cosplay to a higher level with books such as the following:

- *The Steampunk Adventurer's Guide: Contraptions, Creations, and Curiosities Anyone Can Make*, by Thomas Willeford (McGraw-Hill Education TAB, 2013)
- *Steampunk Gear, Gadgets, and Gizmos: A Maker's Guide to Creating Modern Artifacts*, by Thomas Willeford (McGraw-Hill Education TAB, 2011)
- *Make It, Wear It: Wearable Electronics for Makers, Crafters, and Cosplayers*, by Sahrye Cohen and Hal Rodriguez (McGraw-Hill Education TAB, 2018)
- *The Costume Making Guide: Creating Armor and Props for Cosplay*, by Svetlana Quindt (IMPACT Books, 2016)
- *Make: Props and Costume Armor: Create Realistic Science Fiction and Fantasy Weapons, Armor, and Accessories*, by Shawn Thorsson (Maker Media, Inc., 2016)
- *Foamsmith: How to Create Foam Armor Costumes,* by Bill Doran (Punished Props, 2014)

Campers can draw from characters they already love, but you can also help them create new characters based on their own traits and goals. I use a simple worksheet (www.teacherspayteachers.com/Product/Superhero-Application-4699176) to help campers think through what their superpowers might be. They come up with a name and a logo of their own. Then they can use that foundation to develop costume elements that are completely unique.

Masks

COST: Free–$

MAKE TIME: 15–30 minutes

These paper or foam masks are inexpensive and easy to make but provide lots of room for creativity.

Materials

- Chipboard, oak tag, or similar thick paper
- Mask templates
- Colorful cardstock or craft foam sheets
- Scissors, including decorative-edge scissors
- Hole punch
- Glue sticks or white glue
- Markers, colored pencils, ink pads, paint, etc.
- Feathers, rhinestones, stickers, chenille stems, wire, brads, colored foils, etc.
- String, yarn, or ribbon

Process

1. Copy the provided templates or your own template onto paper. Copy at 100 percent size, but also make copies at 90 and 100 percent size so that campers can layer the masks and so that a variety of sizing options is possible (Figures 3.16 and 3.17).

FIGURE 3.16 Masks designed by Jennifer McBride

FIGURE 3.17 Masks designed by Jennifer McBride

2. Use the glue stick to attach the paper to the chipboard or oak tag, and allow it to dry. Cut out your mask templates. Label them as "Template."

NOTE: If you have access to a Silhouette digital cutter, use that to make your templates.

3. Have your campers trace the templates to cardstock or craft foam. Cut out the mask. Use the hole punch to make holes for the eyes for easy cutting. Make sure that your campers have enough of an opening for the eyes so that they can see clearly.
4. Layer and decorate the mask as desired. If needed, give the mask time to dry.
5. Punch holes on either side of the mask. Attach string or ribbon. Tie the string or ribbon at the back of the camper's head, and enjoy the show (Figures 3.18 and 3.19).

FIGURE 3.18

FIGURE 3.19

T-Shirt Capes

COST: Free–$

MAKE TIME: 15–30 minutes

It's easy to collect old T-shirts from campers and friends. Then you can turn them into fun and easy capes for superheroes!

Materials

- Crew neck T-shirts
- Sharp fabric scissors
- Fabric markers, permanent markers, or paint
- Colored felt sheets
- Fabric glue or hot glue
- Feathers, rhinestones, stickers, chenille stems, wire, brads, colored foils, etc.

Process

1. Using sharp scissors, cut off the sleeves, close to the seam. Save the sleeves. This "extra" fabric can be used for decorations or masks (Figure 3.20).

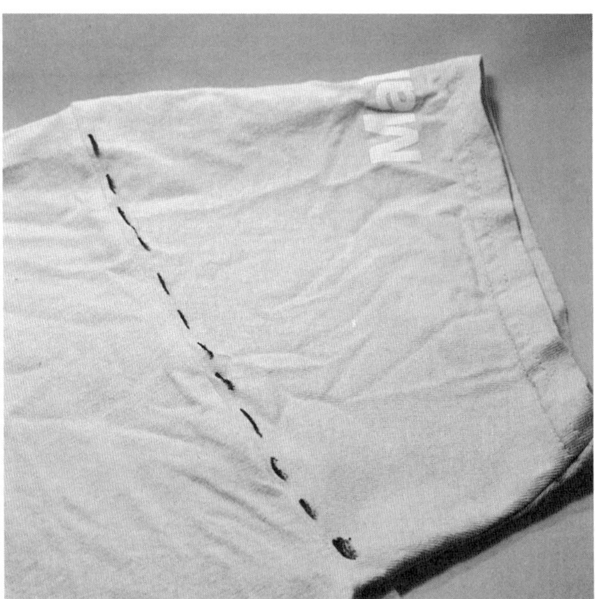

FIGURE 3.20

2. Use sharp scissors to cut from the bottom of the shirt toward the sleeve, just to the side of the seam on the front side of the shirt, if there is one. If there is no seam, use a ruler and pencil to mark the shirt before cutting. Do this on both sides (Figure 3.21).

FIGURE 3.21

3. Decide if you want to use and decorate the blank back of the shirt or preserve the front logo. Cut around the neckline, being careful to preserve the ribbed neckline (Figure 3.22).

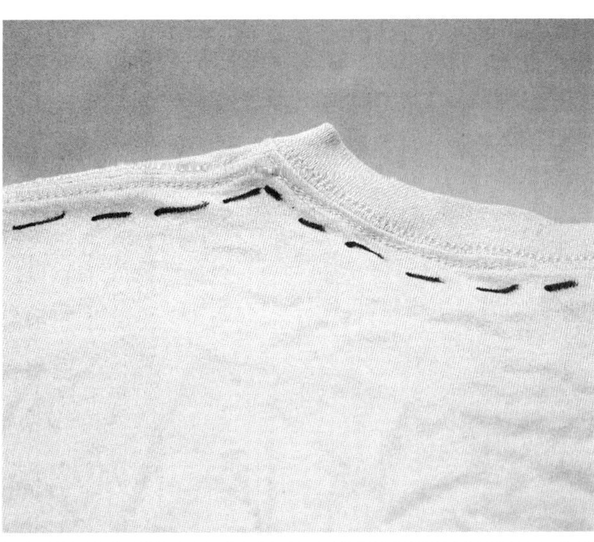

FIGURE 3.22

4. Cut shapes for your superhero logo from the felt. Attach it to the back of the cape using glue. Alternatively, use paint or glue to add a logo.
5. Decorate your cape as desired.
6. To wear the cape, pull the neckline over your head.

Cardboard Swords

COST: Free–$

MAKE TIME: 15–30 minutes

At the heart of the cosplay tradition is the creation of fanciful weapons, often the bigger and more ostentatious the better. Not only can this provide a fun engineering activity for campers, but it also gives them a wonderful opportunity to get creative and use common materials in fun new ways. Swords are often the easiest way to start.

Materials

- Paper and pencil
- Tape measure and ruler
- Cardboard of various types and thicknesses
- Sheets of craft felt, in neutral tones
- Twine and cotton string
- Masking tape
- Aluminum foil
- Brown paper bags, preferably thin ones such as lunch bags
- Hot glue
- White glue
- Paintbrushes
- Scissors
- Box cutters or canary cutters
- Permanent markers

- Acrylic paints
- Feathers, beads, rhinestones, stickers, chenille stems, wire, brads, colored foils, etc.
- Foam balls, table tennis balls, tennis balls, etc.
- Bamboo skewers or wooden dowels

Process

1. Have your campers plan on paper using the tape measure and ruler. They should determine the size and general shape of their creation. It is often best to plan for the hilt and blade separately.
2. Trace out the shape of the blade on a piece of cardboard. Then trace out a few pieces that are smaller, but of similar shape. Cut the pieces using box cutters or canary cutters.
3. Layer the pieces one on top of the other, and hot glue them into place.
4. Use masking tape to tape along the edges, using it to smooth the layers between the pieces and molding them to look like a single piece. Select which side is the "sharp" side, and use tape to compress the edges and make them thinner.
5. Cover the surface with a thin coat of white glue. Carefully wrap the blade in foil, and smooth it out. Let it dry while you work on the hilt.
6. The hilt usually consists of guard, grip, and pommel. The grip is the handle. The guard is perpendicular to the grip and goes between the blade and grip. The pommel is a decorative piece that goes on the end of the sword.
7. Trace the grip on cardboard, again making several pieces that can be layered. Use the same method to create the hilt as you did the sword.
8. Cover the hilt with white glue, and carefully wrap the grip in foil (for a metal hilt) or crumpled, torn paper bags (for leather or wood). Let it dry.
9. For the grip, cut a rectangular or oval shape. Measure the length and width of the blade, and cut a piece out of the center of the guard.
10. If desired, layer several pieces and use masking tape to mold the guard. Cover with a thin layer of white glue, and then cover it with foil or torn paper bags. Let dry.
11. Create a pommel out of a ball, or mold one out of foil, and cover it with masking tape. Cover with a thin layer of white glue, and then cover it with foil or torn paper bags. Let dry.
12. Slide the guard onto the blade, and use hot glue to anchor it in place.
13. Push several skewers or dowels into the base of the blade of the sword. Gently push the hilt onto the skewers, using hot glue to secure the connection. If needed, trim the dowels as needed. Do the same to add the pommel to the end of the sword.
14. Use paint to add shading to the sword. Foam, string, and brads can all add texture. Decorate the sword to give it personality (Figure 3.23).

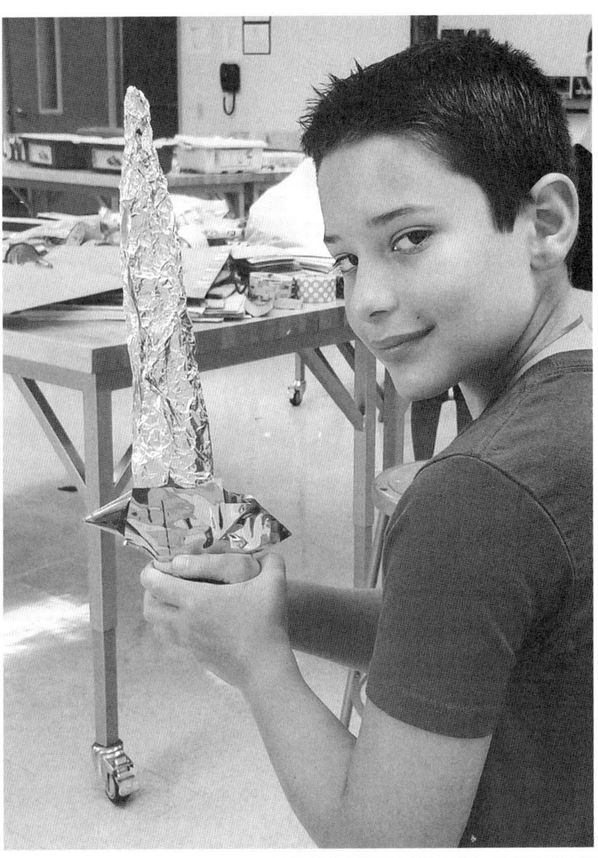

FIGURE 3.23

Take It Further

Try using faux leather strips wrapped around the hilt for an authentic warrior look. Or, instead of a sword, make a staff using toilet paper or paper towel tubes taped and glued together. Collected plastic tubes or rolled plastic sheets can be used with simple, inexpensive tea lights to create a light saber. Get creative, and try unusual materials to create exciting items.

PROJECT 7
3D-Printed Backpack Tags

COST: Free–$

MAKE TIME: 60 minutes

One the easiest and most rewarding introductory 3D-printing projects for campers is making a backpack tag. These tags can be creative, simple to design, and quick to print, so it's easy to get them done and into campers' hands.

For this activity, you'll need access to a computer or tablet with CAD software used to create the print files. You'll also need a way to print your files, either with a 3D printer at camp or by sending files out for printing. If you don't have a printer, consider reaching out to local libraries or makerspaces. You can also try a service such as Shapeways (shapeways.com).

One of the most popular pieces of software is called Tinkercad (tinkercad.com). It is free and works in most web browsers on tablets or computers. Tinkercad allows you to create classes for your campers using an invite code, which will allow you to see and access their work directly and makes it easy for campers under 12 years of age to have an account without lots of extra steps. It also offers tutorials and starter projects. It's intuitive to use, allowing users to pull in shapes and combine them. The one downside is that you must have a strong and reliable internet connection, which can be problematic in some places. Because it's free and widely available, I will be using Tinkercad for this project.

Another popular piece of CAD software is Morphi (morphi.com). Morphi offers an iPad or Google tablet app as well as PC and Mac software for design work. Its cost is minimal, and it offers some really neat features such as the ability to use Augmented Reality (AR) to show your print in real life. If you're working on tablets, it's probably your best option because it's been specifically designed for the hardware. Plus you don't need a network connection to use Morphi, making it perfect for situations where internet access is troublesome.

Another option for schools using the Google Education platform is SketchUp. A free online web browser–based version is available, as are premium versions. SketchUp was originally designed as an architectural tool, so it has a lot of design freedom to draw and manipulate your creations, but it isn't nearly as intuitive as other programs. For example, there are no drag and drop shapes available.

You'll also need to choose a filament. For this kind of project, I usually use a white or natural PLA. It's nontoxic, prints quickly and with nice results, and campers can color their creations with permanent markers or acrylic paints. Of course, if you have access to a variety of colors of PLA, students will certainly enjoy the variety.

One challenge some students face is working in the metric system. You can either use this as an opportunity to increase their understanding or reset the rulers in your software to inches. This may depend on your comfort level and the time you have available. I usually find that reminding campers that 1 inch is just a bit more than 25 millimeters helps (it's actually 25.4 millimeters). For reference, 1 inch is roughly the width of an adult thumb, or the distance from the tip to the first knuckle.

Remind campers frequently to think about the size of the object they are creating. It can be confusing the first time you are working in CAD to keep a sense of scale.

Materials

- Computers or tablets
- Your choice of CAD software, such as Tinkercad
- Slicer software, such as Cura, Slic3r, of MatterControl
- A 3D printer (or printing service)
- PLA or ABS filament
- Metal split key ring

Process

1. Drag a column onto your gridded work plane.
2. Resize the column to make a flat circular tag. Drag the corners (or use the number boxes) to make the column 50 × 50 millimeters. Select the top handle, and pull it down until the height is 3 millimeters (Figure 3.24).
3. Add a rim. Drag a tube onto the work plane. Resize it to be 50 × 50 millimeters and 2 millimeters high. After selecting the shape, click on the arrow to show the shape menu. Adjust the wall thickness to 1.
4. Select the tube. Select the black cone above the shape. Lift the shape up off the work plane. (You can input a distance of 3 millimeters.) Click the work plane, and then select the tube again. Now move it so that it is over the disk you created. Move the view as needed to make sure that the pieces are lined up properly (Figure 3.25).

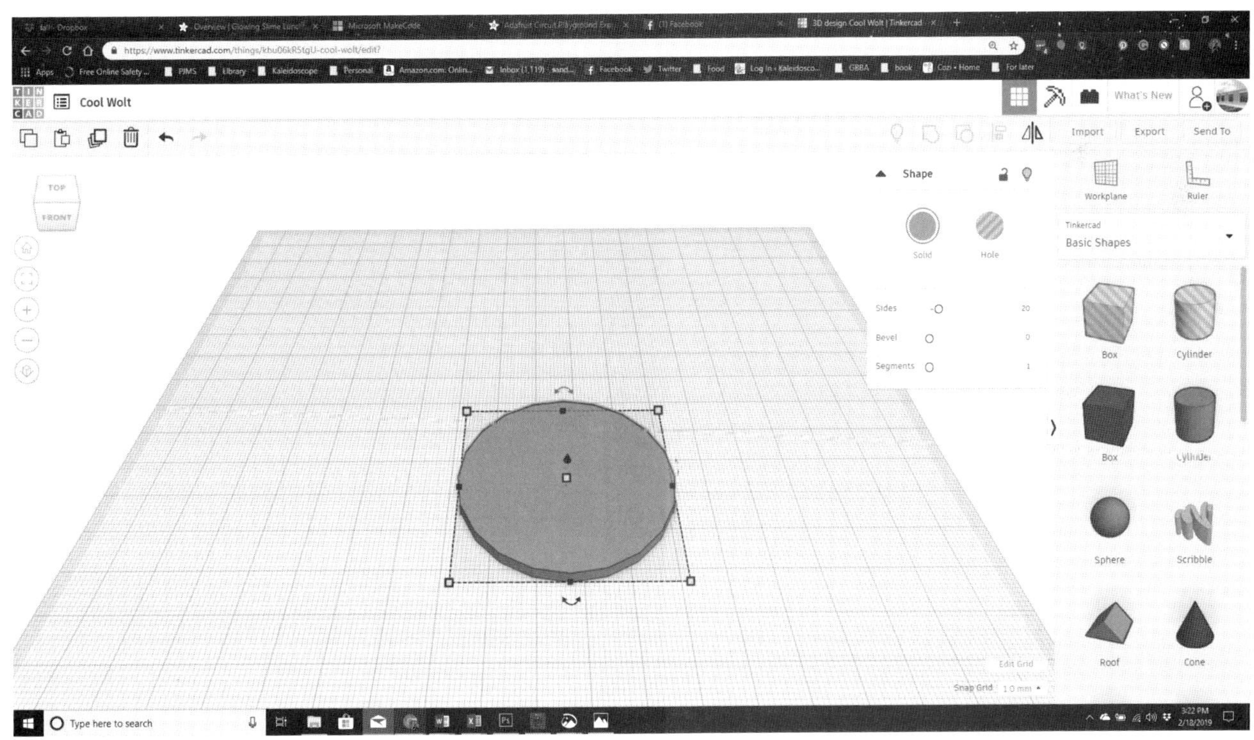

FIGURE 3.24

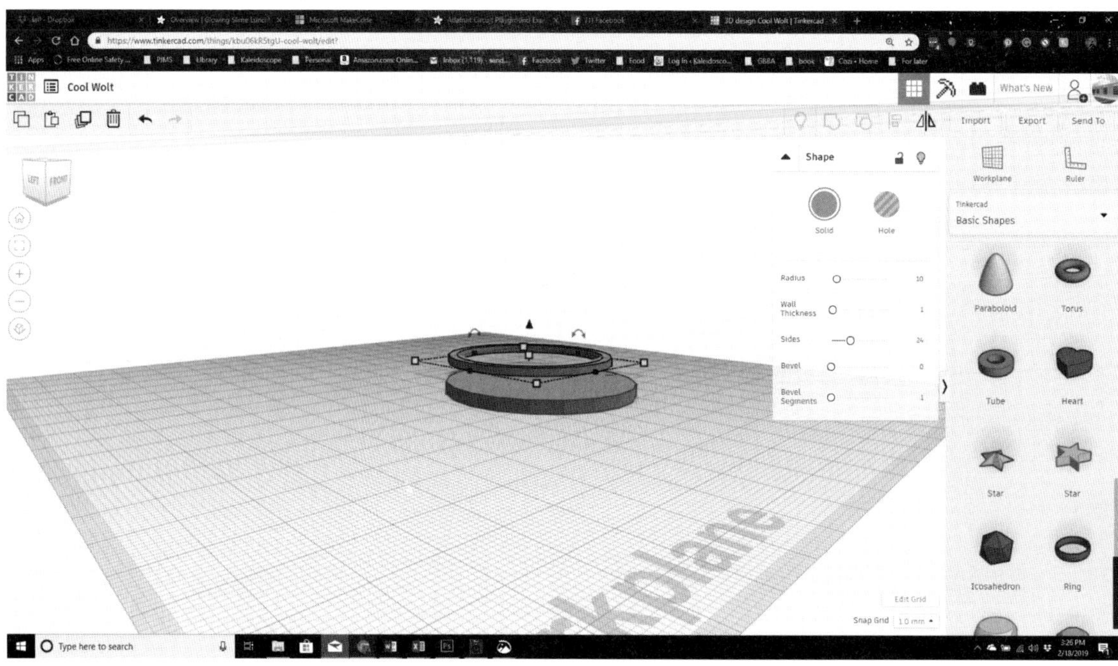

FIGURE 3.25

5. Select both pieces. Click on the group icon to meld the pieces together by selecting the "Group" tool in the upper-right corner or pressing CTRL+G.

6. Drag either a ring or torus onto the work plane. Resize it to be 5 millimeters high with a diameter of 10 millimeters.

7. Move the ring to the top of the disk so that it overlaps. Group the objects (Figure 3.26).

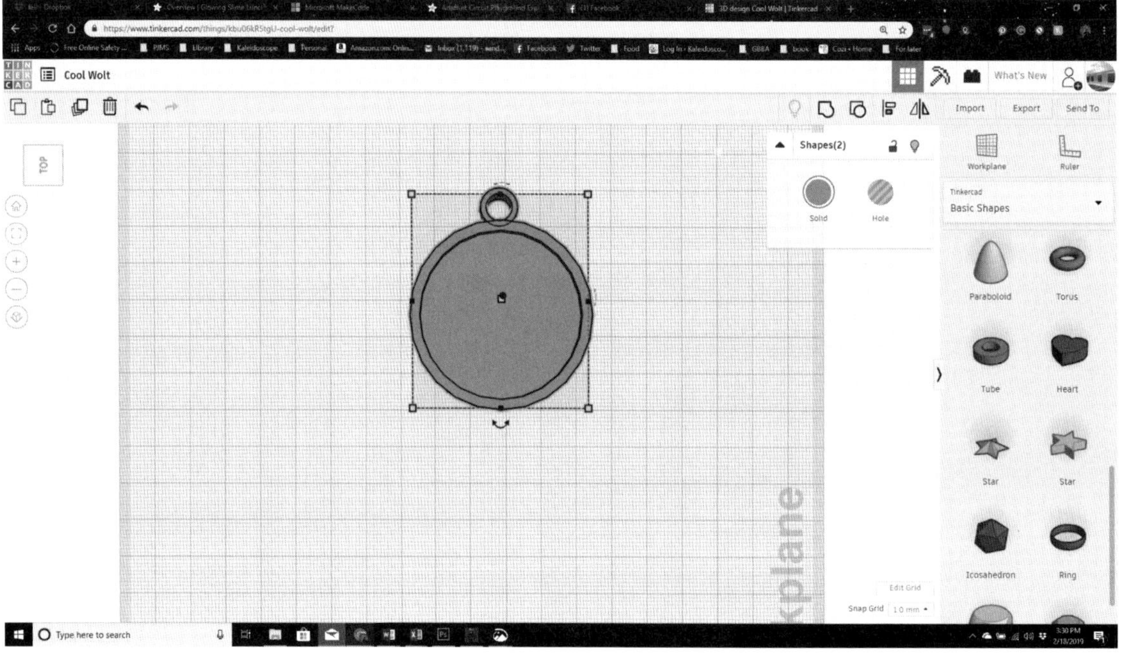

FIGURE 3.26

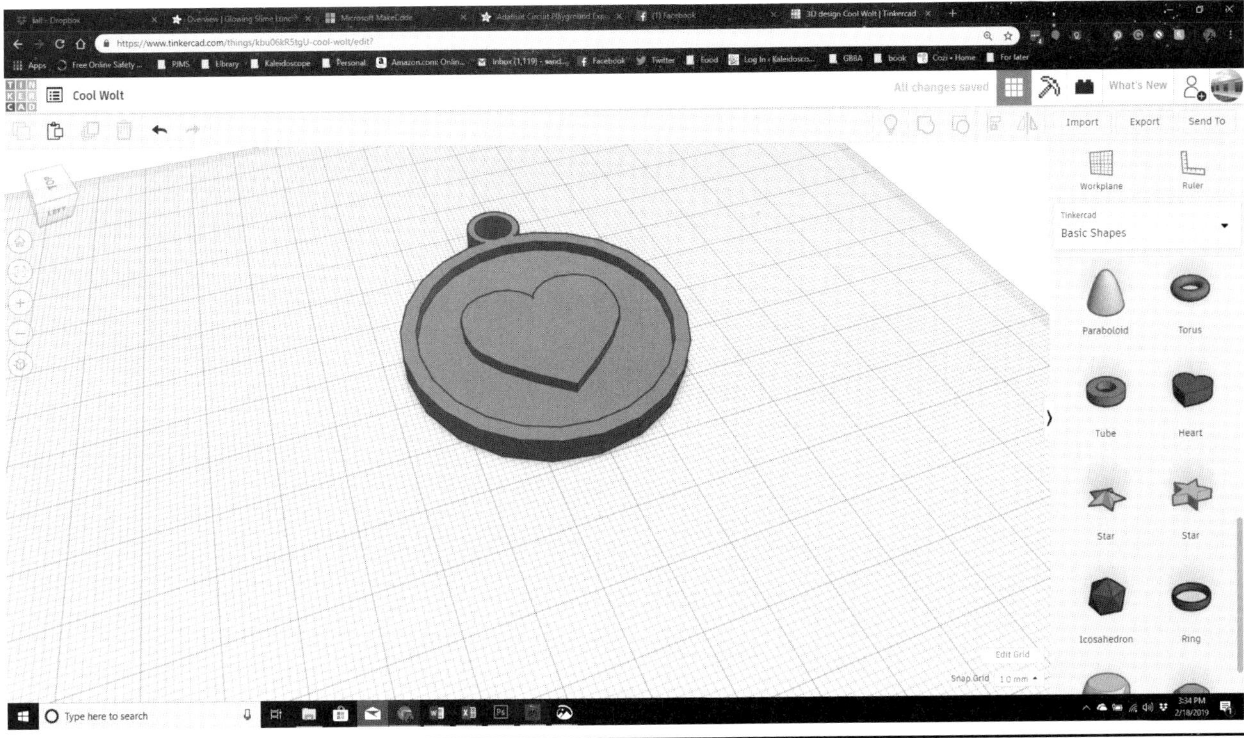

FIGURE 3.27

8. Now it's time to get creative. Add text or symbols as desired. Just keep in mind that 2 inches (50 millimeters) is not a very large space, so designs should be kept simple (Figure 3.27).
9. Download your file from Tinkercad. Import it into your slicer software, and print as appropriate for your 3D printer (Figure 3.28).
10. After it is printed, slide your tag onto the key ring. Slide the key ring onto the end of a zipper or other appropriate location on your backpack.

FIGURE 3.28

Take it Further

You can import .svg images to use in Tinkercad. Simple black and white images work best for this project. You can convert .jpg and .png files in photo editing software or online (see Chapter 6 for details). If you'd like a ready-made Mr. Makey tag for your campers, visit my Thingiverse page to download one at https://www.thingiverse.com/thing:1632408.

CHAPTER 4

Fun and Games

When I was a kid, I used to take apart my toys and put them back together, with varying degrees of success. It drove my mom crazy. I was so curious to understand how they were constructed and how they worked. When toys broke, I wondered if I could fix them, and then I would hobble together new creations from the broken piles.

So I firmly believe that upcycling, hacking, and creating toys are vital parts of childhood. They are also vital maker skills. Whether you are trying to understand a model steam engine or an intricate piece of code, at some point you just have to open things up and look inside to understand how they work.

The projects in this chapter center on using toys in creative ways to remake them into something new and amazing. Some projects are very open-ended. Others have more specific instructions. All of them will be enhanced by the curiosity and creativity of your campers.

PROJECT 1
Lego Labyrinths

COST: Free–$$

MAKE TIME: 30 minutes

This is a fun and easy activity that gives campers a great chance to play with forces and motion. And of course, every kid loves colorful Legos. For this project, you mostly need simple bricks, which are easy to find in thrift stores. If you prefer not to use Legos, try gluing foam craft blocks, sticks, or pasta (ziti, penne) to upcycled carboard.

Materials

- Lego bricks of various sizes
- Lego base plates in various sizes
- Marbles, ⅝ inch
- Graph paper
- Pencil

Process

1. If desired, have campers plan their labyrinths first. Have them count the studs and then use a section of graph paper to represent the base plate. Then fill in blocks to mark the planned route.
2. Campers should plan for an opening that will allow the marble to drop from an upper level to the bottom. They should also plan to make the walls tall enough to keep the marble inside the maze (Figure 4.1).

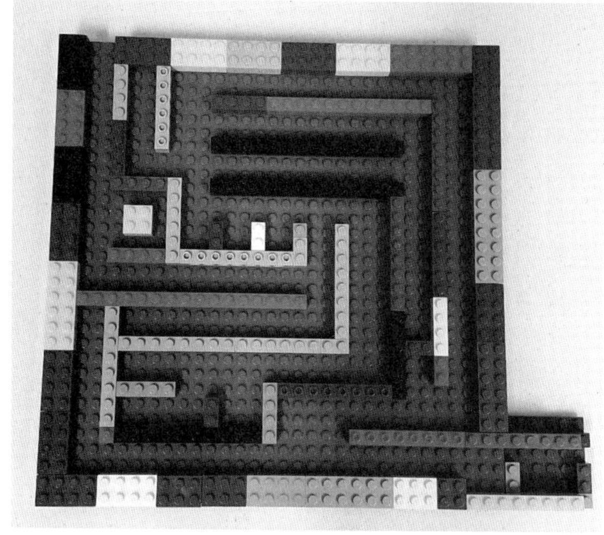

FIGURE 4.1

3. Make sure that campers leave a place for the marble to exit the labyrinth.
4. **Optional activity:** Plan multiple levels, with the largest base plate on the bottom.
5. Once the mazes are built, have campers trade and try each other's creations.

PROJECT 2 Camper "Guess Who?" Board Game

COST: $$

MAKE TIME: 60 minutes

Most of my campers have fallen in love with Fuji Film's new Instax camera and other similar instant cameras. It seems that the modern age has rediscovered the fun of waving a little piece of film around and waiting to see what develops. Who can blame them? These inexpensive cameras make it possible to take lots of photos and share them with your friends. I've used my camera to make instant ID badges, party favors, and light-up paper circuit portraits in the past. The possibilities are pretty endless.

For Maker Camp, I wanted to take it up a notch. My youngest daughter was the one who came up with the perfect project. She loves the game "Guess Who?". This Hasbro classic is a two-player game where players use yes or no questions to identify a mystery character.

Each player has a flip board with 24 different characters. Each character has differing traits such as eye color, hair color, hats, glasses, and so on. Each player picks one character as his or her mystery character. Then taking turns, players may ask questions such as "Does you character have brown hair?" to determine the mystery character's identity. If your opponent says "No" to the preceding question, you'd flip down every character with brown hair. If the player says "Yes," you'd flip down anyone without brown hair. You continue using this process of elimination until you are able to guess who the mystery character is.

With the help of an instant camera, this game is super simple to re-create using the faces of your campers. Be sure to provide "disguises" such as faux glasses, hats, wigs, faux mustaches, and so on. If you have fewer than 24 campers, use your camp leaders or simply have campers take multiple pictures using different disguises. Make sure that you have a solid color wall or background against which to take pictures.

If you don't have an instant camera, you can of course use a smartphone or tablet to take the pictures, import them into Documents on your computer, and print them on a color printer or copier. This may actually be preferable if you want to resize the photos before printing them. A Canon Selphy photo printer or other similar printers are also great for this project, too.

Don't have access to the technology? Many educational companies have "make a face" or "jack o-lantern" stickers that you can order. Just cut index cards to the correct size and have your campers create faces that way.

I offer two methods to make this game below. In the first, you simply use a "Guess Who?" game as a base. This is fine if you don't mind sacrificing a game or just want to include one copy at your game table. The second version, using inexpensive corrugated cardboard, is meant for each camper to be able to take home a game set if he or she wants to.

Modifying a "Guess Who?" Game

This method uses a "Guess Who?" game board. It is modified with your own pictures. There are a number of models out there. The originals from the 1980s are quite a bit different from the ones currently made, so you need to adjust for the game you have. The new version sells for under $15. Older versions can sometimes be found at thrift stores. On eBay, older versions are selling for $15 to $35. If you have an old version in the attic that you are willing to part with, this is probably the easiest way to create a camper "Guess Who?".

Because this version is fairly quick to prepare, consider putting it together after the first camp session. Then set it out for the campers to play the next time you meet. It's a great ice breaker and can be used to help everyone learn one another's names (Figure 4.2).

Materials

- A "Guess Who?" game board by Hasbro (The older versions work better than the newer ones.)
- A DSLR camera, smartphone, or tablet
- A computer with software such as Microsoft Word or Publisher that will allow you to import photos and lay them out as a grid
- A color printer able to print on cardstock
- White cardstock
- Scissors, craft knife, or paper cutter

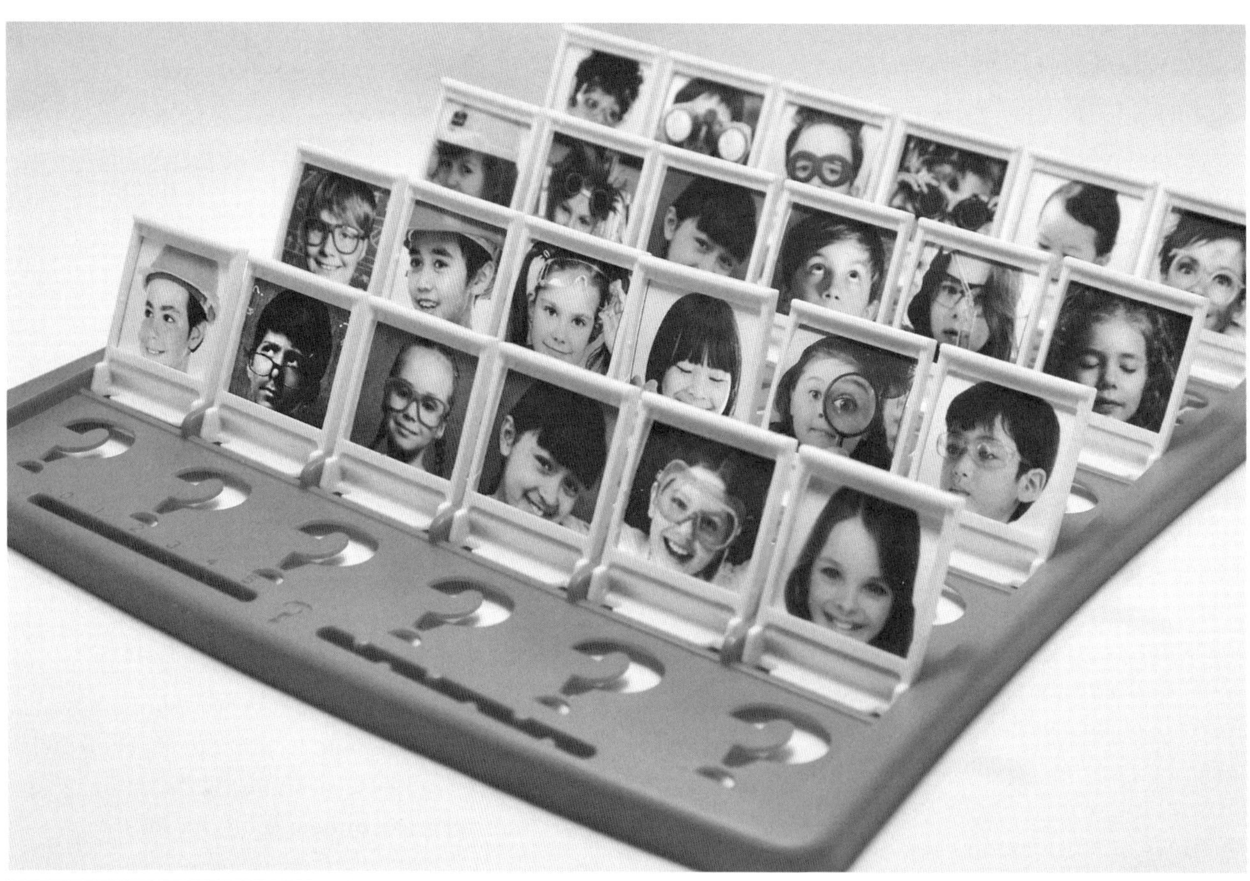

FIGURE 4.2

Process

1. Take photos of all your campers. Preferably, they should be against a white background. If you do not have 24 campers, have some campers take pictures in disguises.
2. Measure the pictures that come with your game. Different game versions vary, so you need to measure your individual game piece photos to make. Generally, the width of the photo will be between 1 and 1¼ inches. The height will be between 1¼ and 1½ inches.
3. Open your editing software. Create a portrait document. Create a table with six columns and four rows. Set the width of your column to the width that you measured. Set the height of your rows to the height you measured. You should be able to fit all 24 photos onto one page.
4. Either add the names in the document or handwrite them later. You can also use a label maker if you prefer.
5. Import all the photos to your computer. Using your software, add each photo to your table, cropping and resizing it as needed. Save the document. Print two color copies onto cardstock. Cut apart the photos.
6. Remove the pictures that come with the game and replace them with your pictures.
7. In your software, change the orientation of your document to landscape. Resize your table to fill the page. Save the document, and print two copies. Center the pictures vertically and horizontally. Resize them as needed. Cut apart the photos. These are your playing decks.
8. Play the game!

Making Your Own Game Board

This version uses inexpensive materials to make the game board. If you photo copy pages of the photos, every camper can make a copy to take home. Although it doesn't have the smooth operation of the first version (above), it does give more opportunity for makers to customize. For example, you can adjust the number of flaps to whatever number of campers you have rather than trying to fit the 24-character model of the original game. Encourage campers to design their own solutions to improve the game (Figure 4.3).

Materials

- A sheet of corrugated cardboard at least 14 × 16 inches
- A Fuji Film Instax instant camera or equivalent
- A ruler
- A pencil
- A craft knife or box cutter
- Adhesive of choice: hot glue, stick glue, white glue, or double-sided tape
- Paint and paint brushes or other items for decorating
- Leather gloves (optional)
- Cutting mat

FIGURE 4.3

Process

1. Take photos of all your campers. Preferably, they should be against a white background. Take two photos of each camper if using an instant camera. If you do not have 24 campers, either plan to adjust your game board or have some campers take pictures in disguise.
2. If you are not using an Instax, following the directions in steps 3 to 5 in the preceding version. Size photos to 2 inches wide by 3½ inches wide to match the Instax size.
3. If you plan to make several copies of the game, make copies of the Instax pictures using a color photo copier. Each game will need two sets.
4. Lay out your board.
   a. Using the ruler, measure a ½-inch border on each size of your game board.
   b. Measure six 2-inch columns separated by at least ¼ inch between columns. Mark the lines.
   c. Measure four 3½-inch rows separated by at least ¼ inch between rows. Mark the lines.
5. With your cardboard on a cutting mat and wearing leather gloves, use the craft knife to carefully cut the two column sides and the bottom (facing you) row, cutting the two long sides and only one short side.
6. Repeat steps 4 and 5 for the second board.

7. If desired, paint one board red and the other blue. Wait for the paint to dry. Paint the other side yellow or white for both boards.
8. Fold each flap up and back, creasing well.
9. Using glue or double-sided tape, attach your photos.
10. Play the game!

Take It Further

Younger campers may enjoy a Maker Camp memory game. Simply print your photos onto cardstock in a grid. Make two copies. Laminate the copies. Cut the pictures apart. Shuffle the cards, and lay them out face down on a table. Take turns turning over a card and attempting to find the matching card.

If your older campers have a favorite fandom (Harry Potter, for example), have them download pictures of characters from the movies and create a "Guess Who?" game based on that.

PROJECT 3 Green-Screened Camp Photos and Videos

COST: $$$

MAKE TIME: 15–30 minutes

Green-screen technology has been used in animation and movies for many years. Now it is so easy and inexpensive to bring the technology to campers that everyone should include it in their programming. If you don't have the budget to purchase a green screen, you can use felt or fabric. Most software will let you adjust the chroma key to work with many different shades of green. The important thing is to pick a color that is unlikely to be worn by campers or found in nature.

As for software, the Do Ink app on the iPad is only $4.99, making it very easy on the budget. I've tried many apps, and Do Ink provides the best results, the most intuitive interface, and the most flexibility of any of the inexpensive or free educational apps. Plus, the results are real time, so campers can see their creations as they make them. The downside is that it is only available for iPads and iPhones.

There are other choices. Green Screen Live Video Recorder offers a free version for both Apple and Android devices, though its functionality is limited. Green Screen Video for Android is very popular, easy to use and inexpensive. Many people use WeVideo for the Chromebook or on laptops to add backgrounds to videos filmed with a green screen, and iMovie is popular for Macs.

To take it to another next level, consider Windows software such as SparkBooth. This multifunctional software lets campers take their own photobooth photos with customized layouts and green-screened choice of backgrounds. I've found it well worth the price, easy to use, and lots of fun for many years with a variety of programs. Combine the software with a laptop and webcam, add an additional photo printer such as the Canon Selphy, and you have a simple photobooth setup that makes every day of camp a ton of fun!

For older campers, consider using a DSLR camera to film video and then software such as Photoshop Premiere to add backdrops. Although this setup takes more time, more money, and involves more of a learning curve to produce results, those results are of higher quality than other options, making it a great opportunity for teen campers interested in diving into the technology.

When it comes to what to film, consider the age of your campers. For elementary school age campers, creating a music video from favorite songs is a simple project. Just play the music, pick a backdrop, and have fun. Older campers may want to try a newscast or product review for a simple project. It also can be fun to re-create a favorite scene from a movie or TV show. Or campers can use Legos or clay or small figures to create animation with the combination of green screening and a stop-motion app.

Materials

- A computer with webcam or tablet
- Photobooth or green-screen software or app (see above)
- Green screen or appropriate fabric
- Various costume elements and props
- Additional green sheets and clothing to create special effects

Process: Photobooth

1. Choose backgrounds in advance. Most green-screen software will let you import multiple backgrounds and swap them easily.
2. Provide photobooth props and costumes or have campers make their own.
3. Have campers take turns snapping their photos. You may want to set some ground rules, such as, "No photobombing," before you begin (Figure 4.4).
4. Print the campers' creations on your photo printer (Figure 4.5).

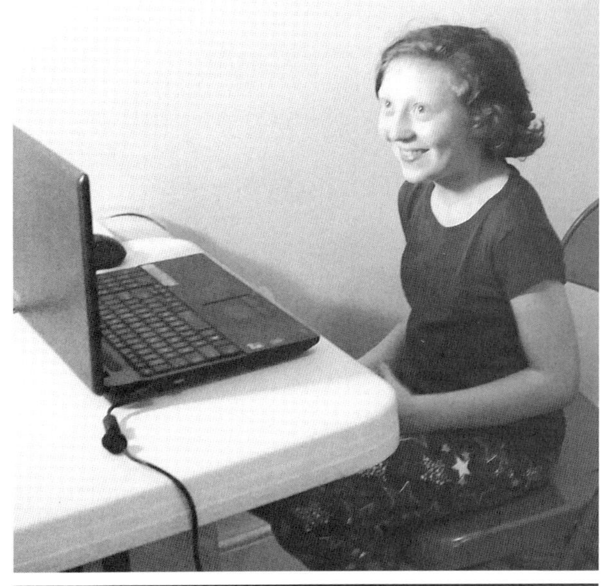

FIGURE 4.4

FIGURE 4.5

Process: Video

1. It can help campers to plan what they will film before they get in front of the camera. Give them a storyboard worksheet to organize their thoughts.
2. Have campers download backgrounds to a designated, easy-to-find directory. This is a great opportunity to talk about trademarked and protected works. Direct campers toward royalty-free photo sites such as Pixabay or Unsplash for their graphics.
3. In a similar way, campers should consider the music they use. The Free Music Archive and PodSafe Audio are great places to start. If campers decide to do a music video, do not share it publicly.
4. Have campers shoot and edit their videos according the instructions provided with the app or software you have selected (Figure 4.6).
5. Share your work on YouTube (if you have permission from all parents to share the work) or on an app such as SeeSaw or Flipgrid, which allows you to control access to the content.

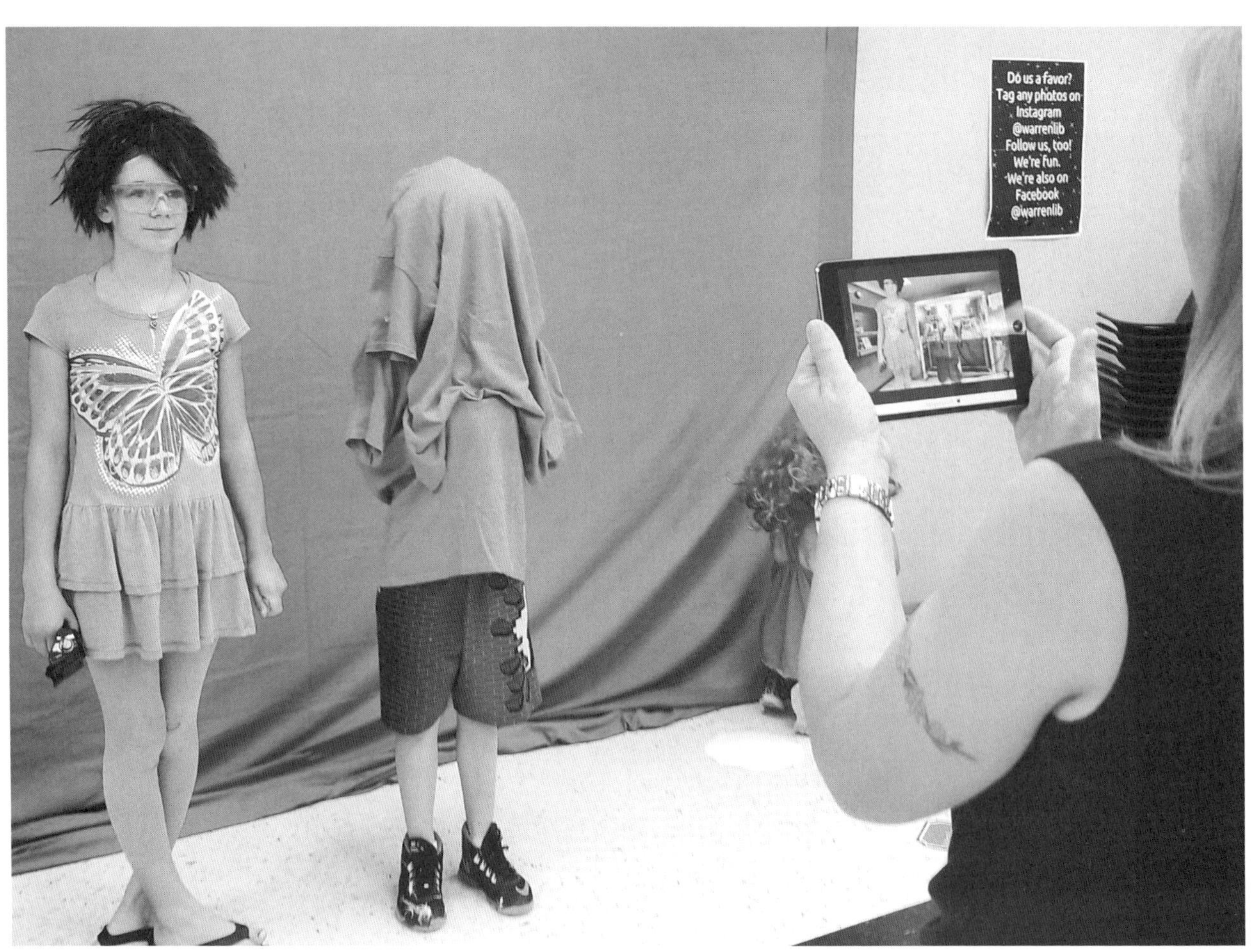

FIGURE 4.6

PROJECT 4
Light-up Letter Home

COST: $

MAKE TIME: 30 minutes

There are many ways to teach the basics of circuits, but many lack—ahem—pizzazz. For Maker Camp, I wanted something that offers creative elements right along with the technical details, which makes paper circuits perfect for learning. Light-up postcards are inexpensive projects with lots of room for adaptation and creativity (Figure 4.7).

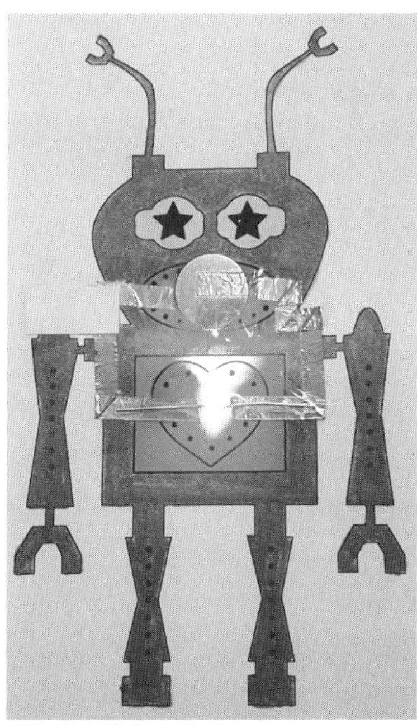

FIGURE 4.7

Before making any light-up cards, you may want to introduce your campers to the basics of how circuits work. There are many good resources online, but here are the basics. A circuit is simply a path for electricity to travel. Usually it includes a power source, a conductive material to allow the electricity to flow, and a load that uses the electricity for some purpose. The circuit may also include resistors, which slow the flow of electricity, and switches, which turn the electricity on and off.

Each load needs a certain amount of electrical force, or voltage (V) to work. The flow of the electricity is called the *current* (I) measured in amperes (or amps). The product of electric force and current is electrical power, measured in watts (think light bulbs). Different materials offer different resistance to the current, creating various conductors and insulators. The resistance to current is measured in ohms (Ω). Voltage, amperage, and resistance are all inversely related, creating the constraints on any circuit. This is expressed by the famous Ohm's law: Resistance = voltage/current, or $\Omega = V/I$. Older campers may enjoy exploring these concepts. Many online interactive labs and games are available online.

Our circuit will use a 3-volt battery as the power source, copper tape as the conductor, and an LED as the load to create light. Depending on the color, LEDs use between 1.8 and 3.3 volts of electricity. Red, orange, and yellow LEDs usually need less voltage than blue, green, and white LEDs. Thus, in general, our 3-volt battery can power one, perhaps two LEDs. The more LEDs on the battery, the more quickly its power will be used.

Most campers begin with a series circuit. In this type of circuit, the voltage flows from the battery, through the load, and back to the battery. You can place multiple loads along the path, but each will use a portion of the voltage. If the voltage needs of the loads, all added up, are more than the power source can provide, the current will stop, and the circuit will not work. These kinds of circuits are easy to make, but obviously, they can be very limited. Also, if any one load stops working, the current cannot continue, so the entire circuit will stop working. This is like old-fashioned Christmas lights, where if one bulb went out, the entire strand didn't work anymore (Figure 4.8).

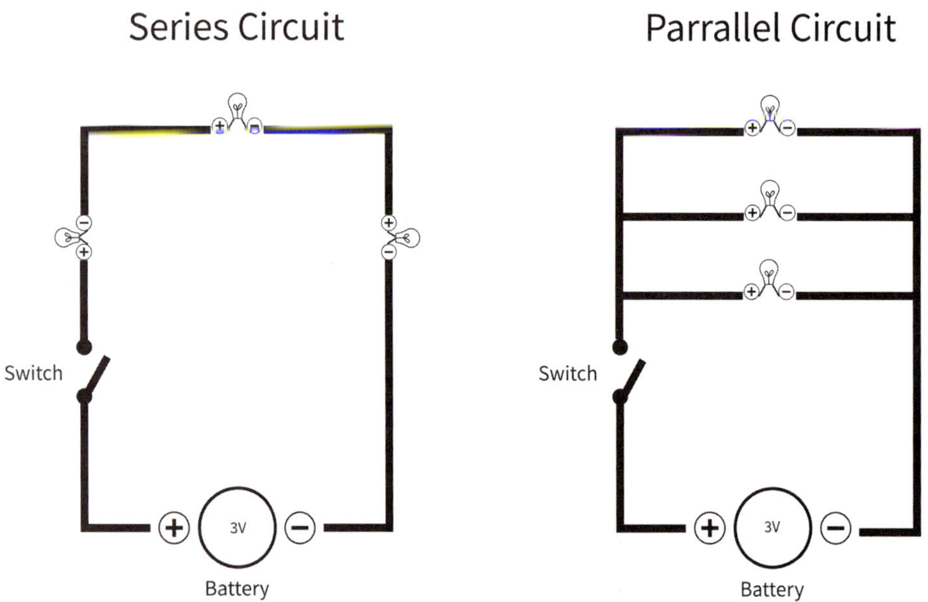

FIGURE 4.8 Diagram by Jennifer McBride

Today, Christmas lights use a parallel circuit. In a parallel circuit, there is a kind of ladder structure to the circuit so that the current flows through each load independently. This means that each load can draw the full voltage of the power source without affecting the other loads. This means that you can power more high-voltage loads on one circuit, although it will drain the battery more quickly. However, if one load stops working, the others keep going. Parallel circuits are more complicated to create, but they give the maker much more flexibility in the design (Figure 4.8).

For this project, I offer a campfire drawing and various templates for circuits, but by no means should this limit you or your campers. Once they have the basics down, your campers can explore making their own pictures and circuits. *The Big Book of Makerspace Projects*, by Colleen Graves, has further ideas, or you can search for ideas on websites such as Makerspaces.com and Chibitronics.

Paper circuits can be finicky. Here are a few of the most common troubleshooting issues:

- **The battery is old or the LED is damaged.** Always test your battery and LED together, as described in the process steps, before taping them into the circuit.
- **There are breaks in the copper tape.** Check carefully to make sure that there are no tears or breaks along the straight lines of tape or on the corners. You need good contact and a solid line of tape to let the electricity flow.
- **There is a short circuit.** The electricity will always take the easiest path to flow from negative to positive. If the copper tape accidentally touches anywhere other than shown in the template, you may have a short circuit that allows the energy to jump across the circuit, bypassing the LED. Most commonly this happens because the camper has used a single piece of copper tape to attach the leads of the LED to the card.
- **The light doesn't work, even though the preceding items are all checked.** The most likely cause here is that either the leads on the LED have been reversed or the battery is connected incorrectly. The flow of electricity

must always go from negative to positive in one direction. Try flipping the battery first. If this doesn't work, try flipping the LED.

Materials

- Template printed on cardstock
- 5-millimeter LEDs in various colors
- 3-volt coin batteries (product numbers 2032 or 2025)
- ¼-inch copper tape with conductive adhesive
- Scissors
- Invisible tape
- Binder clip
- Markers, colored pencils, crayons, watercolor paints, etc.
- Needle-nose pliers (optional)

Process

1. Print the front and back of the sample card on copy paper. If desired, draw your own. Color your project using whatever method you like (Figures 4.9 and 4.10). A printable copy of this template is available at https://www.teacherspayteachers.com/Product/Maker-Camp-Paper-Circuit-4715680.
2. Use the copper tape to trace the circuit path. To do this, you will need to peel off the back of the tape to uncover the adhesive. The

FIGURE 4.9 Illustration by Jennifer McBride

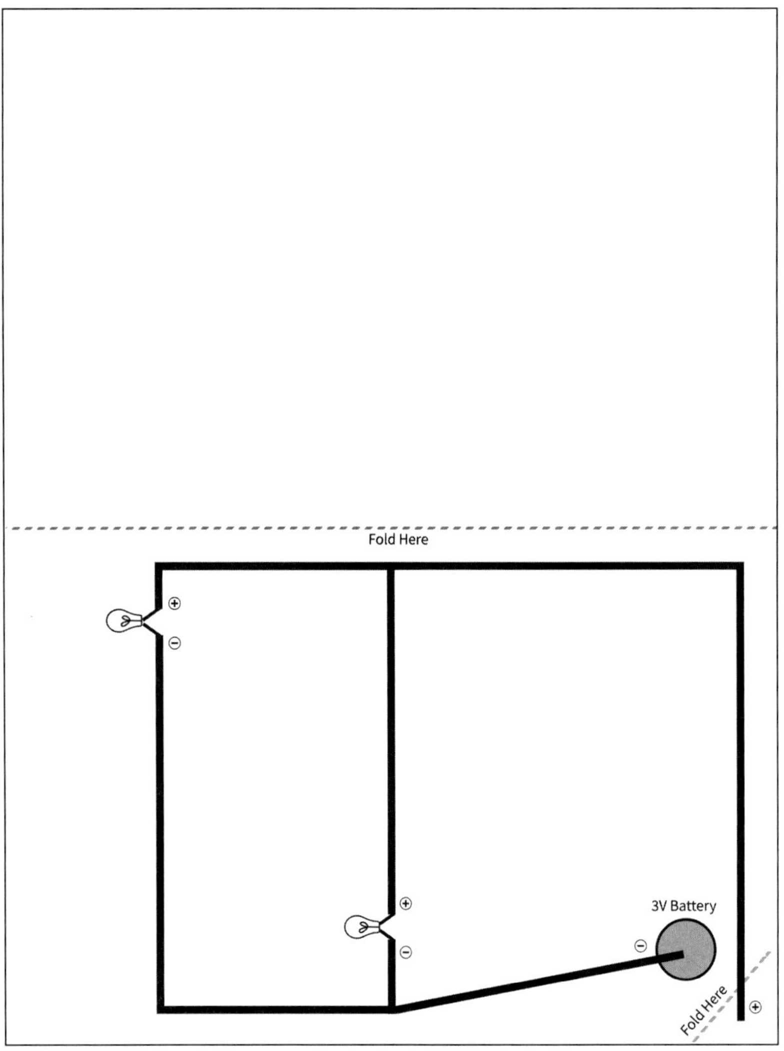

FIGURE 4.10

tape will easily stick to itself, so be sure to only remove the backing a little at a time, pressing the exposed tape to the paper as you go. Try not to create wrinkles or breaks in the tape. Do not cross the tape along lines of your template or across connections. You can handle corners in two ways:

a. Apply the copper tape as one continuous piece rather than separate pieces, gently folding in the corners to make sharp or rounded edges. This is preferred because the conductive adhesive on the back of the tape offers a weaker, less durable connection than the metallic tape itself. It can be a little tricky, but the general idea is to lay down tape to the edge of the corner. Then fold your tape in the opposite direction you want to turn so that the sticky side is face up and flatten. Next, fold the tape back over in the direction of your turn and flatten (Figure 4.11).
b. The folding method can be tricky for little hands and newbies, so sometimes it can be easier to tear or cut your tape to the appropriate lengths for your template and just overlap the tape between sections. It's important

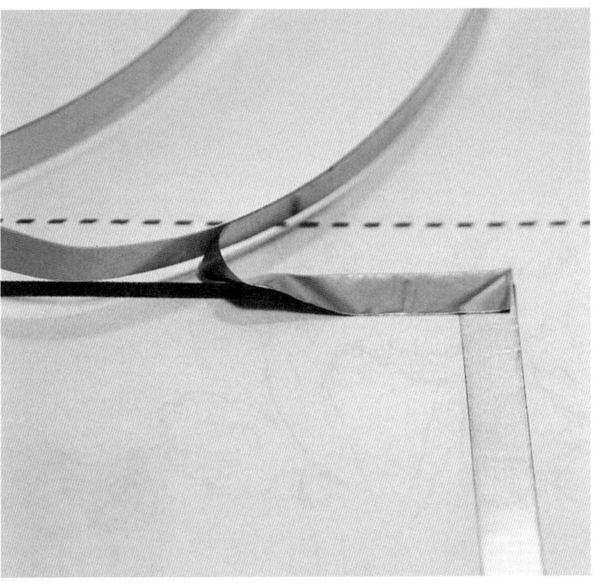

FIGURE 4.11

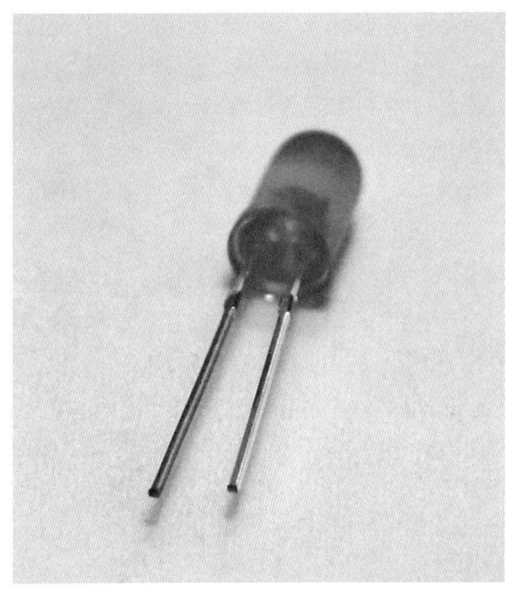

FIGURE 4.12

to make sure that the sections of tape overlap well and have no breaks between them. This method may be easier, but the circuit will be weaker overall, which can cause problems if the camper wants to make more complicated circuits. It should only be considered a starter method, used until the folding skill is learned and perfected.

3. A LED has two legs, or leads. The long one is the positive (+) lead. The short one is the negative (−) lead. It's important to know which is which. It can help to take a permanent marker and color the negative lead black. Test your LED before going any further. Slip the LED over the battery so that the positive lead lies against the positive side of the battery (marked with a +) and the negative lead rests against the other side. The LED should light up (Figure 4.12).

4. Before attaching the LED to the card, you will need to bend the leads. Using your fingers or a pair of needle-nose pliers, *gently* spread and bend the leads of your LED. This will make it easier to attach the LED to your card. Take care not to break the leads. If you like, you can curl the leads into circles. This will give you better contact with the circuit when you attach it. At camp, we call this "bunny ears" (Figure 4.13).

5. Place your LED on the card, and use copper tape to attach it. Note the positive and negative lead locations marked on the circuit diagram. This is really important! Your circuit will not work if the leads are reversed. Electrical energy must flow in the

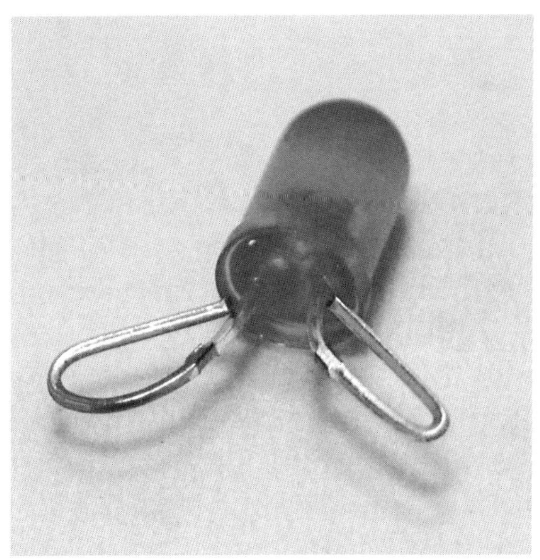

FIGURE 4.13

proper direction, so the leads must match up with the battery for the circuit to work. Make sure that the leads make good contact with the copper tape circuit. Use your fingers to press down hard on the tape. You can also add a piece of clear tape to hold the LED in place.

IMPORTANT: Make sure that the leads are separated and that the copper tape holding the leads down does not overlap negative and positive. This will create a short circuit, and your card will not work (Figure 4.14).

FIGURE 4.14

6. Now it's time to attach the battery. The battery has two sides. The positive side is smooth and marked with a plus sign (+). The negative side is rough and unmarked. Place the coin cell battery with the negative (−) side face down on your circuit. Use clear tape to attach it on the edges if desired, but leave metal exposed. If desired you can fold over a piece of copper tape into a loop, with the sticky side out, and use this to further

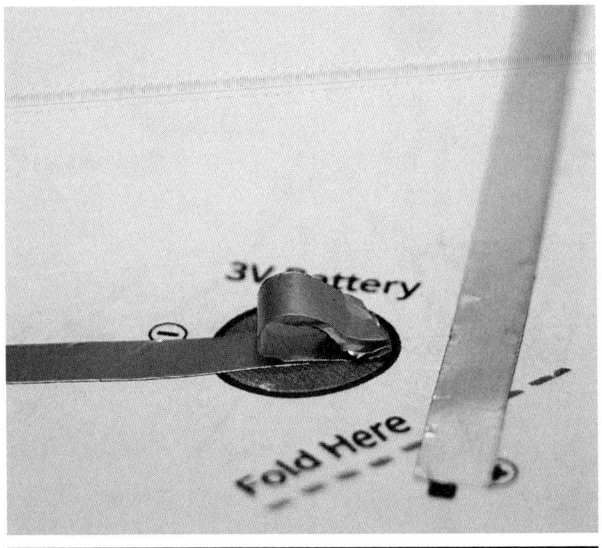

FIGURE 4.15

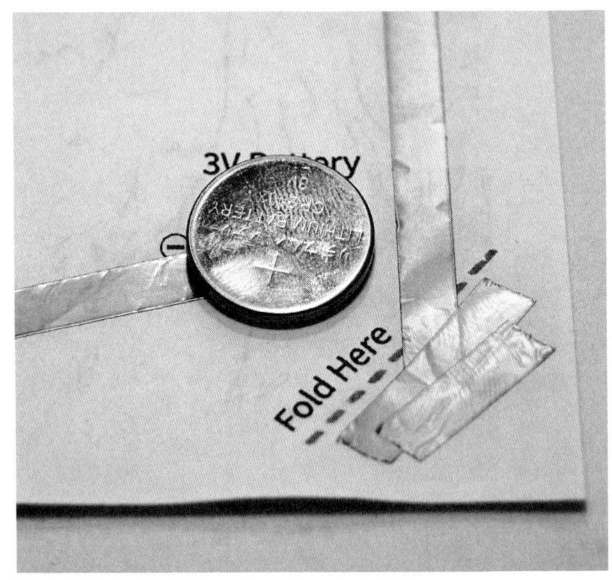

FIGURE 4.16

secure the battery. The goal is to make sure that the battery has good contact with the circuit (Figures 4.15 and 4.16).

7. To turn the circuit on, fold the cardstock at the dotted line, and make sure that the copper tape has good contact with the negative side of the battery. The LED should light up! Use the binder clip to secure the connection (Figure 4.17).

FIGURE 4.17

Take It Further

You can try making the copper tape circuit a creative part of the design, making swirls and words on the front of your card. Or try popping your LEDs through the paper. This is especially helpful if you use cardstock for added weight. Vary your paper choices, using vellum or rice paper. Obviously, this method can be used on surfaces other than paper. Try clear surfaces such as acrylic sheets or Plexiglas. Go three-dimensional (3D) and try adding LEDs to polyvinyl chloride (PVC), wood, or ceramics.

Campers can also add various sensors, such as pressure, light, or sound sensors, to create more complex circuits. Companies such as SparkFun and Chibitronics make sensors designed for such projects, although standard sensors can be adapted for use on paper. Similarly, campers can create all kinds of switches using folded paper, paper clips, brad fasteners, and standard switch components. With a little planning, they can even create interactive displays on paper or connect the paper circuit to a Makey Makey to interact with a computer.

If you want a true hacking experience for your campers, consider harvesting parts from an inexpensive LED tea light. For under $1, you can get a white LED, a battery, and a switch. Check out the Scrappy Circuits website (https://sites.google.com/deweymac.com/scrappy-circuits/home) for details on hacking the light. You can even substitute aluminum foil for copper tape if you want to go completely do-it-yourself (DIY) on your materials.

PROJECT 5
"Smack a S'more"

COST: $$$

MAKE TIME: 60–90 minutes

This game is a fairly simple project that will introduce your campers to the micro:bit. The micro:bit is a programmable open-source microcontroller designed by the British Broadcasting Company (BBC) for use in computer education in the United Kingdom. It is inexpensive and smaller than a credit card, making it very accessible and adaptable. The micro:bit is also very easy to code using either Microsoft's visual MakeCode or JavaScript. As a result, the micro:bit has a ton of free projects and curriculum available online, making it easy for makers to grow their skills and share their creations.

This game is a cross between "Hot Potato" and "Whack a Mole." In the game of "Smack a S'more," you will randomly receive a marshmallow for your s'more. It is your job to pass the marshmallow along to another camper to keep it roasting. Wait too long, and it will burn. If that happens, you are out of the game.

Using the code, the marshmallow will randomly arrive on a micro:bit. If the camper smacks it quickly, it'll pass randomly to another camper. Hesitate, and the marshmallow will burn, knocking you out of the competition. The last one standing wins. The code also keeps score, which can be used to determine who the best marshmallow roasters were. The game can be played alone as well, with one person trying to keep all the marshmallows from burning.

To add a crafty *steam* component, we'll also make simple plush s'mores. These have the added benefit of ensuring that your micro:bit won't be damaged if campers are a little too rough when they smack. My campers had a lot of fun coloring faces on the marshmallows and giving them names. They can even code the micro:bit to reflect the character on the pixeled screen.

Materials

- At least two BBC micro:bits, each with
  - Micro USB cable, 6 inches end to end
  - 2 × AAA battery holder
  - 2 × AAA batteries (make sure they are fresh)
- Leftover corrugated cardboard
- Dark brown craft foam sheets, preferably extra thick (6 millimeters)
- Child's white socks, one per micro:bit
- Fiberfill, either cotton or polyester
- Adhesive foam squares
- Scissors
- Craft knife or cardboard cutter (optional)
- Permanent markers
- Hot glue
- Needle
- White thread
- Paint and paintbrush (optional)

Process: Make the S'mores

These are very easy-to-make and inexpensive soft characters for the game. They really enhance the experience and protect the micro:bits from damage.

1. Using scissors or a cardboard cutter, cut two 3- × 3-inch squares from the cardboard. If desired, paint the cardboard to look like graham crackers and allow the pieces to dry.
2. Using scissors or a craft knife, cut one 2½- × 2½-inch square from the brown craft

foam. If desired, decorate the square to look like chocolate.

3. Push fiberfill into the toe of one of the socks. Stuff the sock until it takes shape but is still squishy. It should make a ball approximately 2½ to 3 inches in diameter.
4. Thread the needle, and tie a large knot in the end of the thread. Sew a simple running stitch around the neck of the sock, just above the level of the fiberfill. Once you have gone all the way around the neck, gently pull on both ends of the thread to cinch the neck of the sock closed. Tie the two ends of the thread together tightly. Use scissors to trim the excess sock from the marshmallow. If desired, use markers to add a face to your marshmallow.
5. Lay a piece of cardboard on the table. Use hot glue to attach the craft foam to the center of the cardboard.
6. Add hot glue to the center of the craft foam. Press the cut edge of the marshmallow sock into the glue, pressing down gently to attach the marshmallow.
7. Use hot glue to attach the second piece of cardboard to the top of the marshmallow sock.
8. Allow the glue to cool and harden before using the s'mores plushies for the game (Figure 4.18).

Process: Code the micro:bit

The micro:bit uses the radio feature of the controller to send messages between micro:bits. Each micro:bit has essentially the same code on it, and each needs to be assigned its own player number. This allows the micro:bits to recognize each other and send the marshmallow around. I suggest using nail polish to paint a number on each micro:bit to help campers keep them straight.

FIGURE 4.18

MS MakeCode is a very easy language to learn. It's a visual drag-and-drop environment similar to Scratch. MakeCode has specialized sets of code for different devices—such as the micro:bit, Circuit Playground, Mindstorms, and more. This makes it very easy to control the sensors onboard.

The coding environment is divided into three areas. On the left is a simulation of the device you are using. This makes it possible to check what your code is doing before you download it. The center features a menu of color-coded function blocks. And the right is the Scripts area, where you actually build the code. On the bottom left is an area to name and save your code as well as the "Download" button. On the bottom right are "Undo" and "Redo" buttons as well as a "Zoom" button.

Because the micro:bit is Bluetooth enabled, you don't need computers to code it. Apps have been developed for iOS and Android. Users can code in the app and use Bluetooth to flash the code onto the micro:bit. This can make it easier for locations that don't have computers available or if you want to work on coding projects outdoors on a nice day.

1. Go to http://makecode.microbit.org, and click on "New Project."
2. Let's start by telling the micro:bit what to do when it starts.
   a. Click on the blue "Basic" menu. Drag the "on start" block into the Scripts area to the right of the screen.
   b. Then click on the pink "Radio" menu, and drag the "radio set group" block into the Scripts area and place it inside the "on start" block. Set the group to "1."
   c. Click on "Basic" again, and select "show icon." Place it into the "on start" block. Set the icon to the frowny face. This will serve as reminder that the micro:bit hasn't been readied yet for play.
3. Now let's get the micro:bit ready for the game.
   a. From the fushia "Input" menu, drag the "on button pressed" block into the Scripts area. Click on the small arrow to show the drop-down menu. Set the block to "A".
   b. Click on the red "Variables" menu. Click on "Make a variable." Create the variable "score," and click "OK." Repeat to make the variables "number," "player," "next," and "smore."
   c. From the "Variables" menu, drag the "set to" block in and place it inside the "on button pressed" block. Set score to "0."
   d. Repeat for "number." Set "number" to the number of micro:bits you are using for the game.
   e. Repeat for "player." Set the "player" for each micro:bit, starting at 1 for the first and going up sequentially. This is the one place the code differs for each micro:bit! No two micro:bits should have the same player number.
   f Repeat for "smore." Set "smore" to "−1." This indicates that you do *not* have a marshmallow right now.
   g. From the "Basic" menu, add the "show string" block to the "on button pushed" block. In the white oval, type "Ready!"
4. To start the game, we need to have a micro:bit send a marshmallow out. We do this by randomly selecting a player and sending a signal using the radio. Once the game starts, this will happen automatically, but we need to actually get the marshmallows roasting at the beginning.
   a. From the fushia "Input" menu, drag the "on button pressed" block into the Scripts area. Click on the small arrow to show the drop-down menu. Set the block to "A+B."
   b. From the "Variables" menu, drag the "set to" block in and place it inside the "on button pressed" block. Set the variable to "next." Click on the light-purple "Math" menu and select the "pick random" oval. Place it into the white oval on the "set next" block. Enter "1" for the first number. For the second number, we want it to go all the way up to the maximum number of players. To do this, go to the "Variables" menu and select the oval that reads "number" (you set this in step 3d). Put the number oval into the second white oval of the "set to" block.
   c. From the "Radio" menu, select the "radio send number" block. Add it to the "on button pressed" block. From the "Variables" menu, drag the "next" oval into the white oval. This will send the random "next" value to the other micro:bits.

d. But wait! What if the random number it picks is your player number? We need to make sure that you don't end up just picking yourself over and over. From the green "Loops" menu, select the "while do" block. Add this to "on button pressed" block in between the "set to" and "radio send number" blocks. From the teal "Logic" menu, select the "=" comparison block. Place it where the "while do" block says "true." Then right-click on the "set to" block we made in step 4b. Select "Duplicate." Now add that copied block into the "while do" block. Now, before sending the next player number, the code will check to see if the next variable matches your player number. If it does, it'll pick a new player.

5. But how will other players get the signal? We need to code the radio receiver.

   a. From the "Radio" menu, select the "on radio received" block and drag it into the Scripts area. Make sure that the drop-down menu shows "receivedNumber."

   b. From the teal "Logic" menu, select the "if then" block and add it inside the "on radio received" block. Drag the "=" block to where the "if then" block says "true." Into the first white oval, place the red "Variable" oval for "receivedNumber." In the second white oval, place the "Variables" oval for "player."

   c. From the "Variables" menu, select the "set to" block and place it in the "if then" block. Set the variable to "smore." From the "Math" menu, drag the "pick random" oval into the white oval. This will set the amount of time a player has to smack his or her s'more. I set it to between 5 and 10 seconds, but you can adjust the time to make the game harder or easier.

6. The pixel screen on the micro:bit will show one of three images during the game. A smiley face if you have no marshmallow, an unburned marshmallow when you receive a marshmallow, and a black marshmallow when you have burned your marshmallow. Let's code those screens now.

   a. From the "Basic" menu, drag a "forever" block into the Scripts area. Then add three "if then" blocks from the "Logic" menu. Place them all inside the "forever" block, one right after the other.

   b. In the first "if then" loop, we will set the screen for times you do not have a marshmallow. This condition occurs any time the "smore" variable is less than 0, such as when we set it to −1 at the beginning (in step 3f).

   c. From the "Logic" menu, drag the "=" block to where the "if then" block says "true." Into the first white oval, place the red "Variable" oval for "smore." In the white oval, enter "0." Set the drop-down menu to "less than" (<) rather than equal.

   d. From the "Basic" menu, add the "show icon" block to the "if then" block. I selected a smiley face, but campers can pick a different icon or even design a new one to represent their team.

   e. The second "if then" block will handle what happens when you get the marshmallow.

   f. Repeat step 6c, but instead of setting the block to "less than," set it to "greater than" (>). When we receive our player number over the radio, we set a random number for "smore" that is above 1. As as long as that number is above 0, you have the marshmallow.

g. From the "Basic" menu, add the "show leds" block into the "if then" block. Create a circular marshmallow outline. This is our unburned marshmallow.

h. Add the "change by" block from the "Variables" menu under the "show leds" block. From the drop-down menu, select "smore." In the white oval, change the number to −1. This will count down each time the "if then" block runs, giving you roughly 1 second for each round. We'll handle how to stop this countdown in a moment.

i. The last "if then" block will handle situations where the player didn't pass the marshmallow in time and it got burned. We'll use that opportunity to place pressure on the other players by adding more marshmallows to the game.

j. Repeat step 6c from above, but instead of setting the block to "less than," leave it set to 0. This will occur if you run out of time trying to pass the marshmallow.

k. From the "Basic" menu, add the "show leds" block into the "if then" block. Create a circular marshmallow with all the LEDs filled in. This is our burned marshmallow.

l. From the "Variables" menu, add the "set to" clock. Select the variable "player" and set the value to 0. This will take a burned marshmallow micro:bit out of the game, because players are only sending marshmallow to players 1 and up.

m. As we did in steps 2a–c, set the "next" variable to a random player number, and send it over the radio. We don't need to add a "while" loop this time because it's impossible to pick our own number anymore. Instead, this loop will just run every second or so and send a random marshmallow out. As more players' marshmallows get burned, more and more marshmallows will go out.

7. So how do we pass the marshmallow? Well this is "Smack a S'more," so it's about to get a little bit crazy.

   a. From the "Input" menu, select the "on" block. Drag it into the Scripts area. From the drop-down menu, select either "3g" or "shake." Either will activate when the micro:bit is smacked. You may need to test this with your plushie s'mores to select the best one for your game. I used the "3g" option, which activates when the micro:bit experiences a force of at least three times the gravitational force of the Earth (3×*g*), roughly the same amount of force as a cough or sneeze.

   b. First, we're going to check for cheating. Some players may think that it's best to just constantly smack their s'more in hopes of never burning the marshmallow. So we're going to code in a point reduction for anyone who does this. If the player has no marshmallow (in other words, his or her "smore" variable is set to "−1"), he or she will lose a point every time he or she smacks the s'more.

   c. Add an "if then" block with a "=" block instead of "true," as you did earlier. Add the variable "smore" to the first white oval, and leave the second white oval as 0. Set the drop-down menu to "less than" (<).

d. Move the "change by" block to inside the "if then" block. Set the drop-down menu to "score" and the number value to "−1."

e. Next, we'll check to see if the player smacked his or her s'more before time ran out.

f. Add another "if then" block under the first. In place of "true," set it up like you did in step 7c, but instead set it to "greater than" (>).

g. Add a "set to" block. Set the drop-down menu to "smore" and the number to "−1." This will place the player into the "safe mode," in which he or she no longer has a marshmallow. The smiley face will appear.

h. Add a "change by" for the "score" variable, and enter "1" so that the score goes up by 1 every time a player passes an unburned marshmallow.

i. Now we need to send the marshmallow to another player. To do this, we replicate the same random selector for the variable "next" that we used at the start of the game. Repeat steps 4a–c, placing it all within the "if then" block.

8. Our last step is just to add the ability to check the score.

   a. From the "Input" menu, add the "on button pressed" block to the Scripts area. Select "B" from the drop-down menu.

   b. Drag the "show number" block from the "Basic" menu, and place it inside the "on button pressed" block. Add the variable "score" in the white oval.

9. That's it! Once the code is adapted for each micro:bit by changing the "player" variable, you're ready to load the code onto the micro:bits (Figure 4.19).

   a. Give your code a name, such as "Smore." Click the "Save" icon (the small floppy disk), and then click "Download." A .hex file will be downloaded to your computer.

   b. Use the USB cable to plug the micro:bit into your computer.

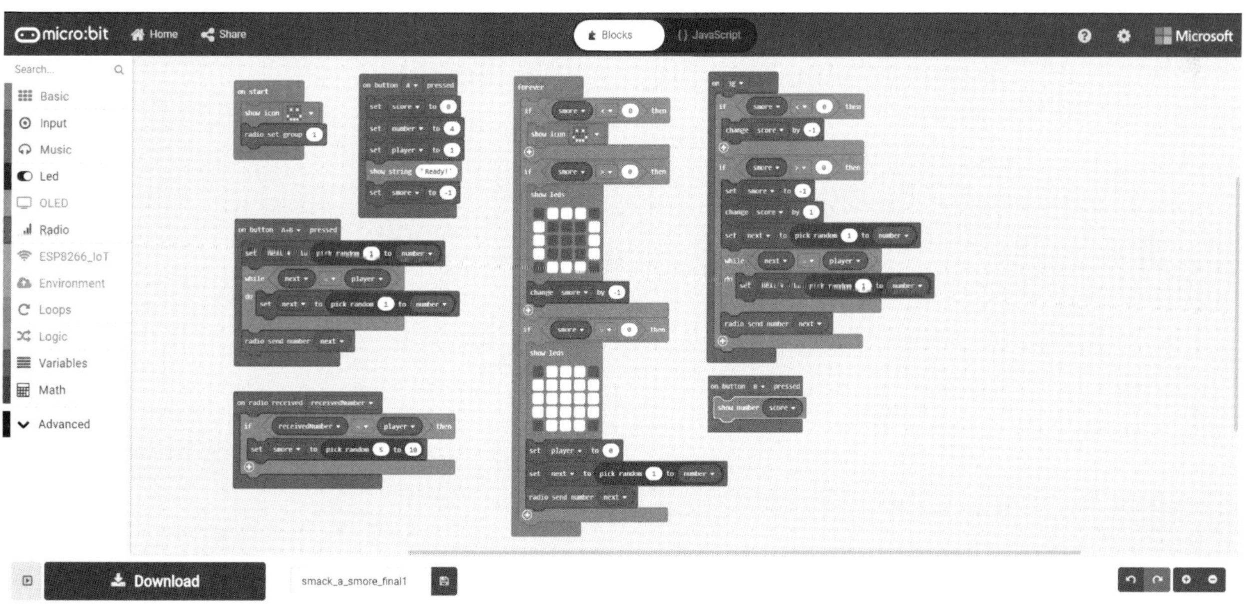

FIGURE 4.19

c. Look for the MICROBIT drive in File Explorer or Finder. Copy the .hex file from your computer onto the MICROBIT drive by dragging it over. The LED on the back of the board will blink, and the file will load. Do not disconnect until it is done.

Process: Play "Smack a S'more" (Multiplayers)

1. Attach each micro:bit to the top of a plushie s'more using an adhesive foam square.
2. Plug in a battery pack to each micro:bit.
3. Arrange players in a circle, perhaps around your LED campfire.
4. Have each player press the "A" button to initialize and reset his or her micro:bit.
5. Then have *one* player press the "A" and "B" buttons at the same time.
6. When a marshmallow appears on the screen, have the player smack or shake the micro:bit until it returns to a smiley face.
7. If the players aren't quick enough, their marshmallows will burn. They are out of the game at that point.
8. The last player to burn his or her marshmallow is the winner.
9. If desired, see who successfully passes the most marshmallows by pressing the "B" button.

Process: Play "Smack a S'more" (Single Player)

1. Attach each micro:bit to the top of a plushie s'more using an adhesive foam square.
2. Plug in a battery pack to each micro:bit.
3. Arrange s'mores in a line in front of the player.
4. Press the "A" button on each micro:bit to initialize and reset the device.
5. Press the "A" and "B" buttons at the same time on *one* micro:bit.
6. When a marshmallow appears on the screen, have the camper smack or shake the micro:bit until it returns to a smiley face.
7. If the camper isn't quick enough, the marshmallow will burn.
8. When the last marshmallow burns, have the camper press the "B" button on the micro:bit to tally his or her score.

NOTE: Download a complete copy of the code at https://makecode.microbit.org/_gCqgFqchvfce.

PROJECT 6
Makey Makey Dance Off

COST: $$

MAKE TIME: 60–90 minutes

This project is more than just coding—although the skills campers will develop using Scratch 3.0 are important. It's combined with engineering a controller for the game that uses a Makey Makey. The Makey Makey is a low-cost tool that allows you to connect any conductive materials to a computer and use it as you would the arrows, ENTER, and click. Here we'll use it to have a large game mat on which campers can "dance" to control the computer.

Even if you don't have a Makey Makey, this is a fun game to play. The idea is simple: the computer randomly selects different dance moves for the onscreen character. You have to hit the key (or match it on the dance mat) to score points. Move too slowly, and you lose a point. You have 60 seconds to score as many points as you can.

Once campers understand the basics of this code, they can modify and remix it. Perhaps they'd like to change the music or the character. Maybe they'd like to create levels, when the dancer moves more and more quickly for each round. Or they could add more complex moves and combinations.

If you have a webcam, campers can even include themselves in the game, taking pictures in various dance positions and importing them as graphics for your program. Got musicians at your camp? Why not create your own sound loop for the game. The possibilities for customization are endless.

Coding the Game

This project uses Scratch 3.0 because this easy-to-learn, drag-and-drop coding language works on both computers and tablets such as the iPad, making it wonderfully mobile. Scratch is completely free to use. All work is saved automatically on the website when Scratch is used online or can be saved locally using both the online or offline version. Coding works by placing various puzzle-like blocks together.

Materials

- Computer or tablet connected to the Internet

Process

1. Go to https://scratch.mit.edu/. If needed, create an account. Click "Create" to get started on your code.
2. The Scratch coding screen is divided into four major sections.
    a. The Stage is located in the upper-right section of the screen. This is where your game or animation is acted out.
    b. Below this is the Sprites area. This shows the characters, or sprites, that are used in your code. Clicking on each sprite allows you to control what they do and how they look. This area also allows you to add new sprites or delete ones you no longer want. You can also change the backdrops shown on the Stage.

c. To the far left is the "Block Palette." It has three tabs: "Code," "Costumes," and "Sounds." Under the default "Code" tab, you can select various coding blocks to use in your program. Blocks are color coordinated and grouped by purpose. The "Costumes" tab has tools that let you edit or create sprites and backdrops. The "Sounds" tab lets you select, edit, and create sounds to use in your program.

d. In the center is the Scripts area. Here you will build the code for each sprite. When you select a sprite and then drag blocks into the Scripts area, you are telling that sprite what to do in the Stage.

3. Let's start by creating sprites and background.

   a. In the bottom right-hand corner, under the "Stage" menu, click on the "Choose a Backdrop" button. Select the magnifying glass, and choose a background you like.

   b. In the Sprites area, click on the cat. Remove the cat by clicking on the "X" in the upper right-hand corner.

   c. At the bottom of the "Sprite" window, there is a "Choose a Sprite" button. Click it, and select the magnifying glass. A new screen will appear. From the menu along the top of the screen, select "Dance." Hover over each dancer, and select one that you like. Click on your chosen sprite. When it appears on your Stage, drag it to the center, or use the controls above the Sprites area to set the X and Y coordinates to 0, 0. Name this sprite "Dancer."

   d. Next, we need to make the four targets for our dancer, each of which will represent a different dance move. Click on the "Choose a Sprite" icon again, but this time select the paintbrush. This will bring up the Paint Editor. Select the circle or square tool. Draw a shape in the center of the editing panel. Use the paint bucket "Fill" tool to fill it with a color of your choice. Name this "costume1."

   e. The Sprite menu is to the left. Right-click on the new sprite you just made, and choose "duplicate." Name the new sprite "costume2." Use "Fill" to make this costume white. Then click on the "Code" tab to return your Scripts area.

   f. At the top of the Sprites area, there are options for your new sprite. Name your target "Block1" by typing it into the bubble next to "Sprite." If the sprite is too large for your Stage, use the "Size" option to make it smaller. Drag this sprite all the way to the left of the Stage.

   g. Right-click on "Block1," and select "duplicate" to make three more blocks. For each, click on the "Costumes" tab to change the color of "costume1." Drag the blocks so that they form a line along the front of the Stage (Figure 4.20).

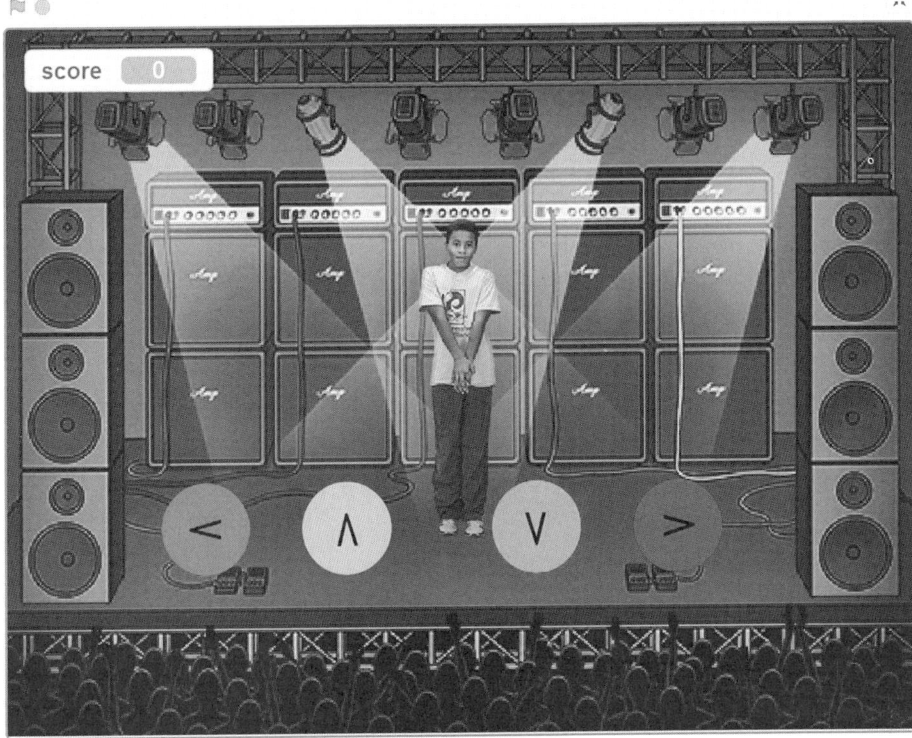

FIGURE 4.20

4. Click on the "Dancer" sprite. We'll start by creating some variables and setting them up. A variable is a placeholder for some value that can be changed and used by the code. We'll use them for such things as keeping score, setting a timer, and tracking which block our dancer is going to.

   a. Click on the orange "Variables" menu. Click on "Make a Variable." Make the following variables: "game timer," "move timer," "score," "block," and "last block."

   b. From the yellow "Events" menu, drag the "when [green flag] clicked" block onto the Scripts area. Anything attached to this block will be triggered when the green flag at the top of the Stage is clicked.

   c. Go back to the "Variables" screen, and select the "set to" block. Drag it into the Scripts area, and place it under the "when [green flag] clicked" block. Clicking the small downward arrow will show a drop-down menu. Select "score" from the menu. In the white oval, enter "0" if it isn't already set to 0.

   d. Repeat the preceding step for each of the other variables, adding them under the previous block. All the variables should be set to 0 except "game timer." Set this to the number of seconds you'd like your game to play for. I set it to 60, for a 1-minute game.

5. Now let's get our dancer set up for the start of the game. Got to the purple "Looks" menu, and drag the "switch costume to" into the Scripts area. Add it under the "when [green flag] clicked" block. Select a costumer for your dancer to start with. Also, from the "Looks" menu, add "go to layer," and select "front" so that your dancer is always in front of other sprites. From the blue "Motion" menu, select the "go to x y" block and add it to the rest. Enter "0" and "0" into the white ovals so that the dancer will always start in the center of the Stage.
6. Let's get each of the blocks set up too. Click on "Block1." Add the "when [green flag] clicked," "switch costume to," and "go to x y" blocks to the Scripts area. Switch the costume to "costume1" for each. For the X and Y coordinates, check the area above the sprites to see where it is located on the stage. Repeat for each block.
7. Return to the "Dancer" code. We need a way to start the game and control how long it lasts.
   a. From the "Looks" menu, select the "say for" block and add it to the "when [green flag] clicked" block. Instead of "Hello," type in "Tap space to start!" Enter the "3" in the second oval.
   b. From the "Events" menu, select the "when space key pressed" block. Add this to the Scripts area. We'll build a new block of code under it.
   c. From the light-orange "Control" menu, select the "repeat" block. Add it under the "when space key pressed" block. From the "Variables" menu, pull the oval "game timer" block into the Scripts area, and place it into the white oval on the "repeat" block.
   d. From the "Variables" menu, drag the "change by" block in and place it inside the "repeat" block. From the drop-down menu, select "game timer." In the white oval, enter the number "−1."
   e. From the "Control" menu, add the "wait second" to the "repeat" block. Enter "1" in the white oval. Now the "repeat" block will start at the number you set for "game timer" and count down by 1 each second, creating the timer for the game.
   f. When the game ends, we'll want something to happen. Add the "switch costume to" and "go to x y" blocks under the "repeat" block. Set the costume to your choice. Set the X and Y coordinates to 0, 0. From the "Looks" menu, add the "say" block. Instead of "Hello," type in "Game over!"
8. You're doing great! Now we need to actually make the dancing happen. This is a really long piece of code, but much of it is repeated. So it's not very difficult to code. First, we're going to randomly select one of the four blocks for our dancer to go to. Then we'll check to make sure that it's not the same as the previous block. And lastly, we'll code for the dancer to move to each block and strike a pose.
   a. Add another "when space key pressed" block to the Scripts area.
   b. From the "Control" menu, drag in a "repeat until" block. From the green "Operators" menu, select the "=" triangular-shaped block. Add this to the "repeat until" block. Into the first white oval, add the oval "game timer" from the "Variables" menu. Leave the second white oval as "0." This will stop

the dancing when the game timer runs out. It should read "repeat until game timer = 0."

c. From the "Variables" menu, place the "set to" block inside the "repeat until" block. Set the variable to "block." From the "Operators" menu, drag the "pick random" oval block into the Scripts area, and place it into the white oval. Set the second number to "4." This will randomly select a number between 1 and 4. If you have more than four blocks, increase the number. It should read "set block to pick random 1 to 4."

d. From the "Control" menu, select the "if then else" block, and add it into the "repeat until" block. Add a green "=" block. In the first oval, add the variable "block." In the second, add the variable "last block." This will compare the block we just selected with the last block used. We'll assign a variable to the "last block" later.

e. Into the first section of the "if then else" block, we're going to add the same "set to" block as we did above (step 8c), with the variable set as "block" and choosing a random number from 1 to 4.

f. From the "Control" menu, select an "if then" block, and place it into the second opening of the "if then else" block (under "else"). Add the "=" block. Set the expression to read "block = 1." Everything in the "if then" block will be for your "Block1" sprite.

g. From the "Events" menu, select the "broadcast" block, and add it to your "if then" block. Fuse the drop-down message to create "new message." Make that message "1," and select it.

h. From the "Looks" menu, add the "switch costume to" block. Select a costume for "Block1."

i. From the "Motion" menu, select the "go to" block, and add it. From the drop-down menu, select "Block1." This will move the dancer to the block and change his or her costume so that he or she looks to be moving.

j. From the "Control" menu, select the "wait" block. Set it to "2 seconds." You can make this time longer or shorter to change the difficulty of the game.

k. Create additional "if then" blocks for each block. Place each one under the previous "if then" block within the "if then else" block. Change the "broadcast" message for each one, and make sure that the dancer moves to the correct block. Change the costumes for each block. The code will look at the random value of "block" and then select the correct "if then" statement to follow (Figure 4.21).

9. Now we need to add code to each block to allow players to score when they hit the right key.

a. From the "Events" menu, select "when key pressed," and add it the Scripts area. From the drop-down menu, set the key to "left arrow."

b. From the "Looks" menu, add "switch costume to" under the "when key pressed" block. Set it to "costume1." This will set the costume back to its initial state once the button is pressed.

c. Add an "if then" block.

d. From the "Operators" menu, select the "and" block. Then select the "greater than" (>) block, and place it into the first position on the "and" block.

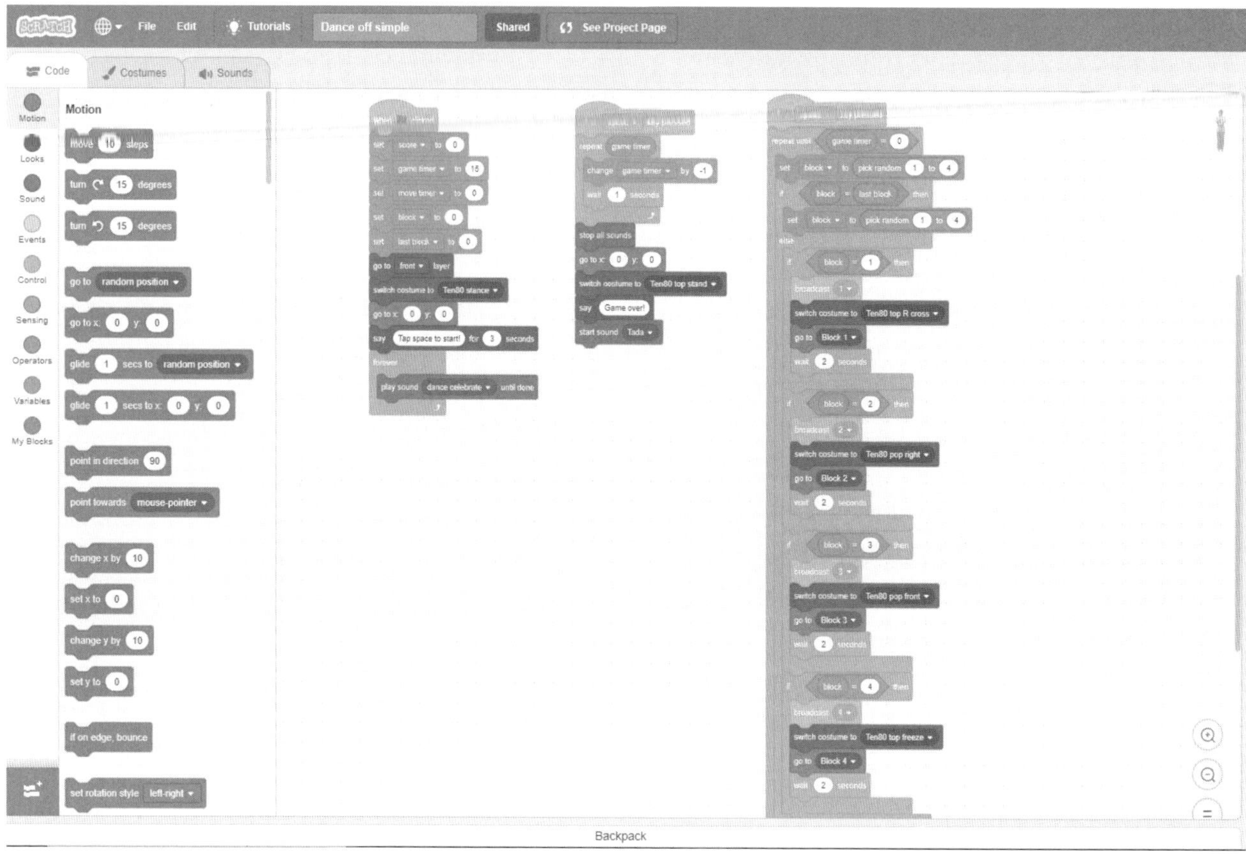

FIGURE 4.21

e. Drag the variable oval "move timer" into the first oval, and set the second to "0."

f. From the light-blue "Sensing" menu, select the "touching?" block, and add it to the second position on the "and" block. From the drop-down menu, set it to "Dancer."

g. This combination will have the code check that when the left arrow is pressed, the player will only score a point if the timer for each move has not run out and the dancer is still at the block.

h. Inside the "if then" block, add the "change by" variable block. Set the variable to "score" and the number to "1."

10. The final piece to code is to make the block "active" only while the dancer is on it.

    a. From the "Events" menu, select the "when I receive" block, and add it to the Scripts menu. From the drop-down menu, select the message "1."

    b. From the "Variables" menu, add the "set to" block. Use the drop-down to select "last block." Add the variable "block" to the white oval. The block should read "set last block to block." This will help the game make sure that the same steps aren't repeated one right after another.

    c. Add another "set to" block. Set the variable to "move timer," and type "2" into the white oval. You can change this number, but it needs to match the time you selected for the dancer. This will control how long the block is active. Making the time shorter makes the game harder. More time makes it easier.

d. From the "Looks" menu, add "switch costume to" under the "when key pressed" block. Set it to "costume2." This will make the block change to white while it is active.

e. From the "Control" menu, add a "repeat until" block. Add the operator "=." Set the variable "move timer" to "0."

f. Place the "change by" block into the "repeat until" block. Set the variable to "move timer" and the number to "−1."

g. Add a "wait" block to the "repeat until" block, and set it to "1."

h. After the "repeat until" block, add a "switch costume to" block, and set it to "costume1" (Figure 4.22).

11. You'll need to duplicate this code for each of your blocks. You can select a group of blocks and use CTRL+C to copy it. Click on the sprite where you want to paste the code, and then use CTRL+V to paste. You will need to change the "when key pressed" blocks for each block. Block 2 should be "up arrow." Block 3 should be "down arrow." Block 4 should be "left arrow." You must also change the "when I receive" block to the correct number for each block.

12. All that's left now is to add a bit of music. Go back to the Dancer sprite. At the end of the "when [green flag] clicked" block add a "forever" loop from the "Control" menu. Inside it, add the "play sound until done" block from the pink "Sounds" menu. To select a sound, click on the "Sounds" tab at the top of the window. Click on the speaker icon at the bottom of the sound editing

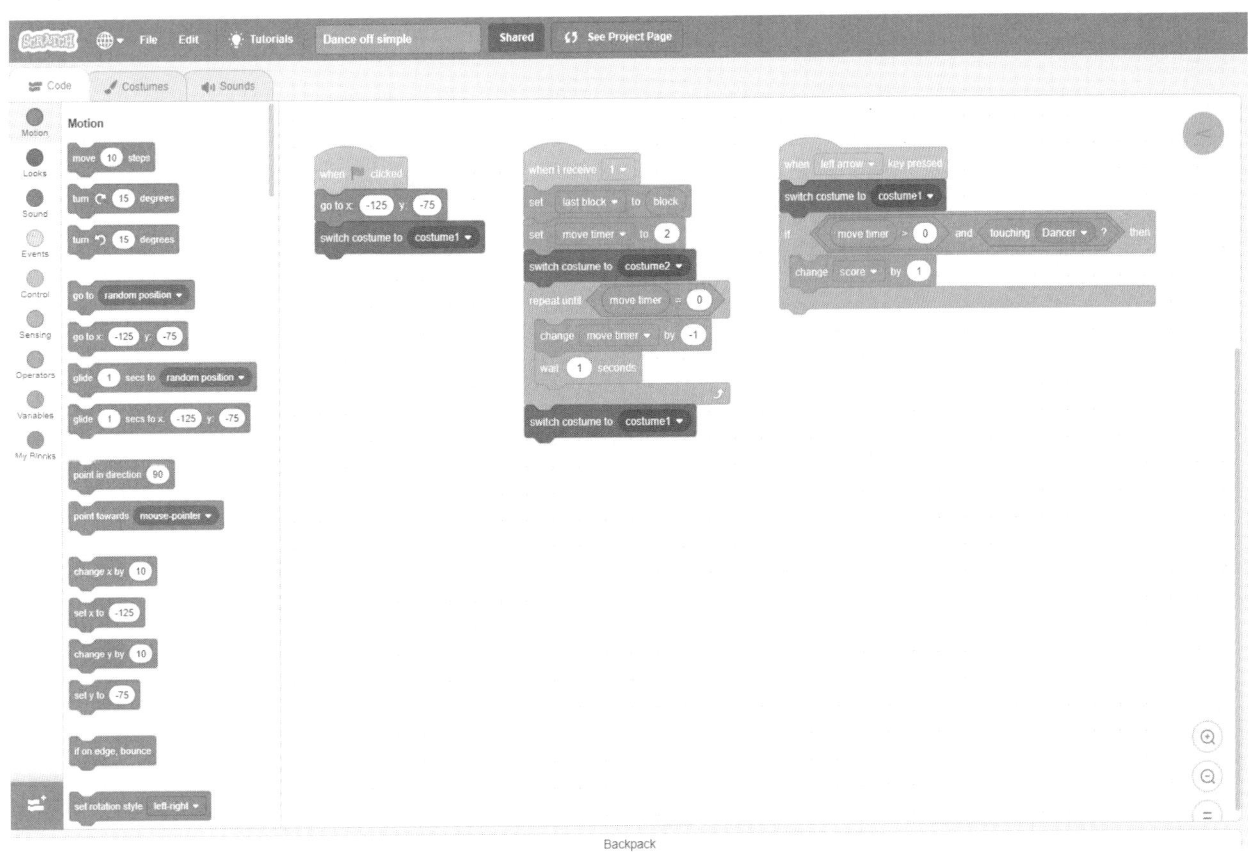

FIGURE 4.22

page, and select the magnifying glass. Select "Loops," and pick your favorite.

13. Give your game a name, and click "Save" under "File" at the top of the window. Test your game, and make adjustments as needed. If you want to, click on "Share your work." Make sure that you add instructions about how to play to your project page.

Take It Further

You can add additional backdrops to work as an introduction and "Game Over" screens. Use the Paint Editing screen to add text as needed. Select a backdrop, and then select the "Code" tab. You can add code blocks to the Scripts area for the stage as well as for sprites. You can have the backdrop change when a player interacts with the game, for example, by clicking on the screen, or when something happens in the game, such as the timer running out. The "Events" menu has a "when backdrop switches to" that you can use to activate other parts of the code, such as the dancer or blocks. You may need to use the "hide" and "show" blocks from the "Looks" menu for your sprites. Play with them, and see what you can create!

You can check and test the sample code for this project at https://scratch.mit.edu/projects/280889743/. For examples of additional elements you can add, see https://scratch.mit.edu/projects/280381887/.

Creating the Dance Pad

The dance pad is made of cardboard with four large "blocks" for campers to tap with their feet. Campers should be encouraged to make the buttons on their dance pattern similar to the blocks they made in their Scratch code so that it is easier for players to follow. You may also want to consider putting rubber under the dance pad to prevent slipping.

Aluminum foil is used to conduct the signal to the Makey Makey. If you prefer to use conductive paint or copper tape, that will also work. The Makey Makey works by creating a circuit between the computer and objects campers create using conductive materials. The Makey Makey board connects to the computer through the USB port. This also powers the Makey Makey. The board connects to the objects created through alligator clips leads. One clip goes back to ground, which grounds the board. The other clip attaches on one end to the object you have created and on the other end to locations on the Makey Makey that link to various keys on a keyboard.

For the computer to receive the signal that a specific key has been pressed, the circuit for that key must be completed. This happens when the lead out for the key connects with the ground, allowing energy to pass from the board to the object and back to the board. We'll do this by having two aluminum foil–covered pieces of cardboard touch.

The Makey Makey can handle up to six keys: left arrow, right arrow, up arrow, down arrow, space, and click. We'll use most of them to control our program. The code already uses those keys as input.

Materials

- Makey Makey with USB cable
- At least 6 alligator clip leads
- Corrugated cardboard
- Aluminum foil
- Glue stick or white glue
- Paint or colored paper
- Permanent markers
- Double-sided adhesive foam squares
- Ruler or tape measure
- Cardboard cutter, box cutter, or scissors
- Duct tape
- A standard 9-inch pie plate (optional)

Process

1. Cut a piece of cardboard that is about 4 feet wide by 1 foot long. Cover the front of the cardboard with glue. Lay aluminum foil over the wet glue, and smooth it to cover the surface, wrapping the foil around the back of the cardboard. Use duct tape to secure the foil on the back of the cardboard. This is the dance pad (Figure 4.23).

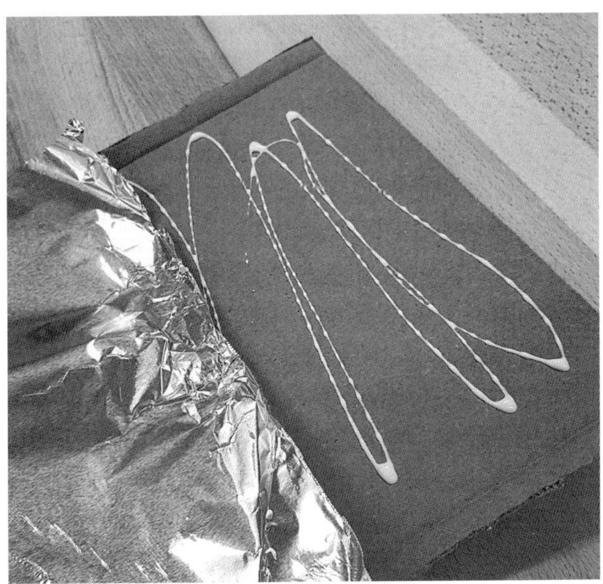

FIGURE 4.23

2. Using the 9-inch pie plate, trace four round shapes onto the cardboard. These will be our blocks. (You can, of course, trace any shape you prefer, but the size should be between 9 and 10 inches in diameter.) Cut out your shapes.
3. Spread glue on one side of each block. Cover it with foil, and smooth the foil out. Trim excess foil. Paint the other side with colors to make your Scratch program, or use colored paper cut to the correct shape and glue it to the front of the dance block (Figure 4.24).

FIGURE 4.24

4. Allow all the pieces to dry.
5. Lay your circular pieces on the foil-covered base, centering them vertically and spacing them out across the cardboard. Lightly mark where each will go using a pencil or marker.
6. On the underside of each block, place squares of double-sided foam around the rim or edges (Figure 4.25).

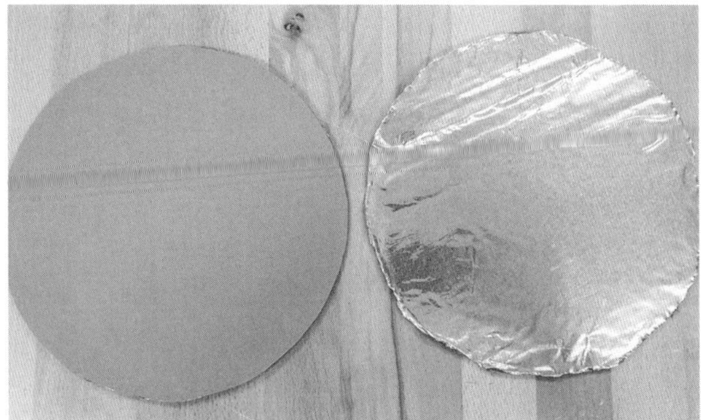

FIGURE 4.25

7. Remove the backing from the foam, and place the blocks on the base.
8. Connect your alligator clips leads. Attach ground to the dance pad base. Attach "Space" to the space block. Attach the "left arrow" to block 1, the "up arrow" to block 2, the "down arrow" to block 3, and the "right arrow" to block 4. Use a little duct tape on the underside of the blocks to hold the alligator clips in place and to reduce the chance of an accidental connection between the clip and ground (Figure 4.26).

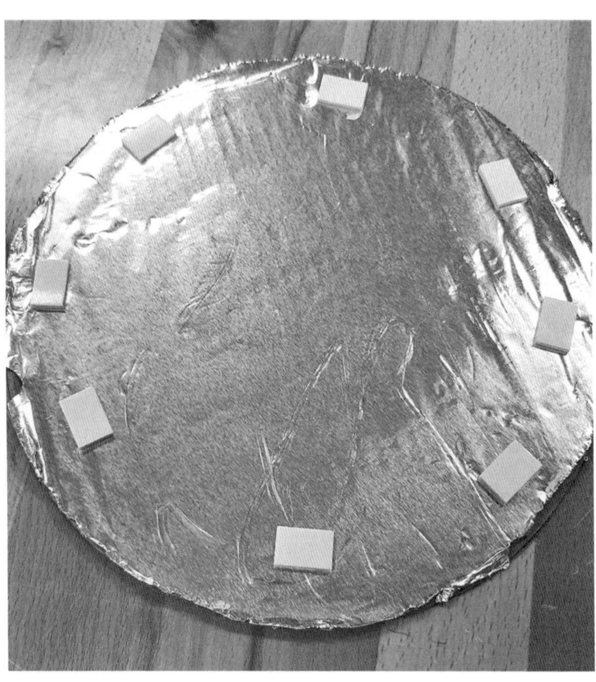

FIGURE 4.26

9. Plug the Makey Makey into the USB port of your computer.
10. Test the dance pad by pressing on each block and ensuring that the key is activated as planned. Make adjustments as needed (Figure 4.27).

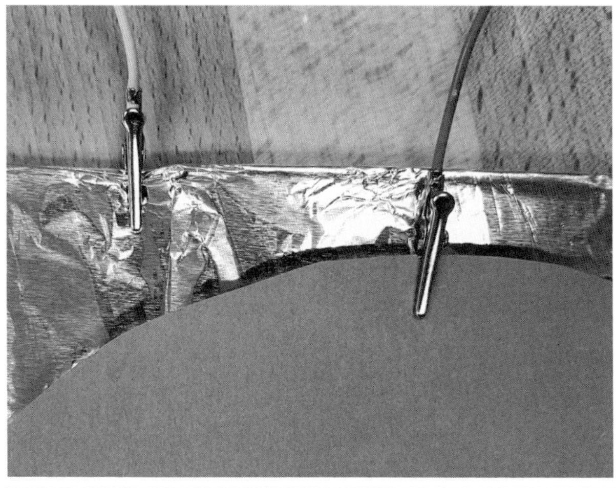

FIGURE 4.27

11. Decorate your dance pad to match your game.
12. Play!

PROJECT 7
Nerf Wars

COST: $$–$$$

MAKE TIME: 1–3 hours, plus drying time for paint and adhesives

Did you know that nerfing is a thing? It is. Nerf is a popular brand of air-powered foam dart blaster presently owned by Hasbro. Both kids and adults alike love everything Nerf related. Invented in 1969, Nerf literally has dozens of different blasters on the market at any given time, and fans of the toy are devoted. I've seen folks host all-out paintball-style raids as well as more gentle Steampunk duels all with Nerf weapons.

There are lots of examples online of some amazing transformations folks have made to Nerf guns, customizing them into works of art and engineering marvels. Because they are relatively inexpensive, easy to find, and easy to modify, Nerf weapons are a summertime standard for my Maker Camps (and my home).

Your first decision when it comes to modifying your Nerf weapon is whether you want the changes to be cosmetic or functional. Cosmetic enhancements are generally easier, and you don't usually need to take the gun apart. Most of the materials you need can be found at the local craft store. Functional enhancements can be more complex and need a wider range of tools and materials. Modifications may include barrel replacement, spring replacement, and modifications to increase the air pressure, such as removing the air restrictor. For these types of adjustments, the gun will need to be opened, taken apart, modified, and rebuilt. Functional enhancements can be tricky, vary widely by the type of gun you're using, and can take quite a bit of time, so I only recommend those activities for older, motivated campers. If you plan to do both cosmetic and functional enhancements, do the functional modifications first. Let the age and interest of your campers guide you on what kinds of enhancements you want to tackle.

If you are providing the Nerf blasters, and I suggest you provide the Nerf blasters to campers, to keep things equitable.. It's okay to use off-brand toys. Campers can transfer the knowledge and skills they build working on a small, inexpensive model to their own gun at home. If the cost is an issue, you may also have success if you ask families to donate old, unused blasters.

Cosmetic Enhancements

The most obvious cosmetic change you can make is to repaint your Nerf gun. The first step is to use a spray paint primer that can adhere to plastic. I personally recommend Rustoleum's 2X Ultra Cover Primer Spray. For the best coverage, you should take the gun apart to paint it, but when I'm doing many items for camp, I don't bother taking them apart, and the results are still quite good. Make sure that you plan enough time to let the spray paint cure before decorating any further.

For the custom paint job, I like craft acrylics. They're inexpensive, easy to blend, and offer lots of colors. Grab some metallic and glow-in-the dark paints in addition to standard colors. Remind campers to apply thin coats so that they dry quickly. A blow dryer on cool can help with the drying process. Obviously, you'll want lots of paint brush sizes. Cheap foam brushes are great for base coats, whereas small artist brushes help with details. Offer a variety of rags, paper towels, and sponges to help create texture with the paint. There are lots of YouTube videos out there that offer tutorials on creating faux finishes with paint.

Depending on the stories your campers are trying to tell, they may want to distress their blaster. A craft heat gun can be used to bubble the paint. Selective use of sandpaper, steel wool, or a nail file used on dried paint can add lots of depth to a piece. Don't be afraid to rough up your weapon either, creating gouges, chips, indentations, and grooves representing faux battle damage. Go ahead and rub in sand or dirt into the cracks, and then brush away the excess. Coffee grounds and leftover tea work well for staining a piece. Sprinkle these substances over the piece, and let them sit for a while. Details such as these are lots of fun to create.

Once your paint job is complete, consider sealing it with a clear acrylic spray in either matte or gloss finish, as appropriate. One or two coats will greatly enhance the durability of the paint, preserving all the hard work that went into it.

When it comes to materials to decorate the Nerf weapons, remind campers that you are creating illusions. You'd be surprised how different things look after a good coat of paint. Brass fasteners make fantastic rivets. Paper bags, when lightly dampened, crumbled, and flattened, look like old leather once painted. First aid gauze adds an amazing burlap-like texture, whereas common cotton string can look like intricate metal work. Craft foam can be cut and layered to create amazing effects. Masking tape takes paint well and can be wrapped in layers and shaped. Even common cardboard can be used to create futuristic or Old West holsters. And don't forget craft standards such as feathers and plastic gemstones. Bits of chain and old jewelry are great as well.

NOTE: A quick note on safety: blasters and plastic toy guns sold in stores look unrealistic for a good reason. Please discourage campers who want to make their creations look too much like a "real gun." Unfortunately, this can meet with tragic results. Keep your creations colorful.

Materials

- Nerf blasters, or the equivalent
- Spray paint primer meant for plastics
- Acrylic paints and paint brushes in a variety of colors
- Clear acrylic spray paint sealant
- Materials to distress the blaster, such as sandpaper
- Materials to decorate the blaster, as described above
- Hot glue or superglue
- Scissors and awls

Process

1. Start by giving all the blasters at least one coat of primer. Use a neutral color such as gray or brown rather than white or black, which can be tougher to cover in the next step (Figure 4.28). Priming should be done outside or in a well-ventilated area while the temperature is between 50 and 90 degrees F and humidity is below 65 percent. Place blasters on scrap cardboard, newspaper, or drop cloth to protect your surfaces. Let your blasters dry 24 hours before moving onto the next step.
2. Add any items to your plan that are used for texture, such as craft foam, string, or fabric. Use hot glue or superglue to secure items, and let the glue dry completely (Figures 4.29, 4.30, and 4.31).

FIGURE 4.28

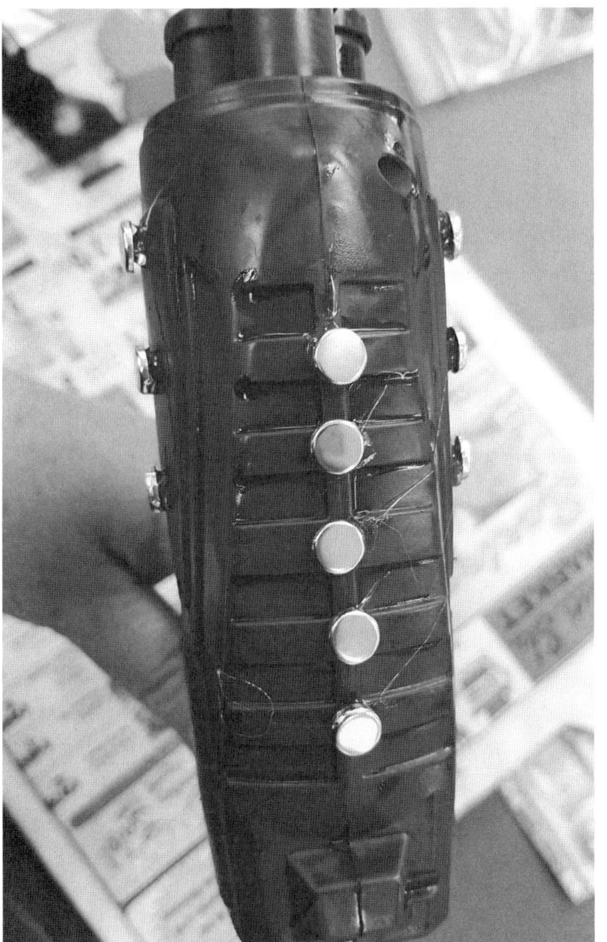

FIGURE 4.29

FIGURE 4.30

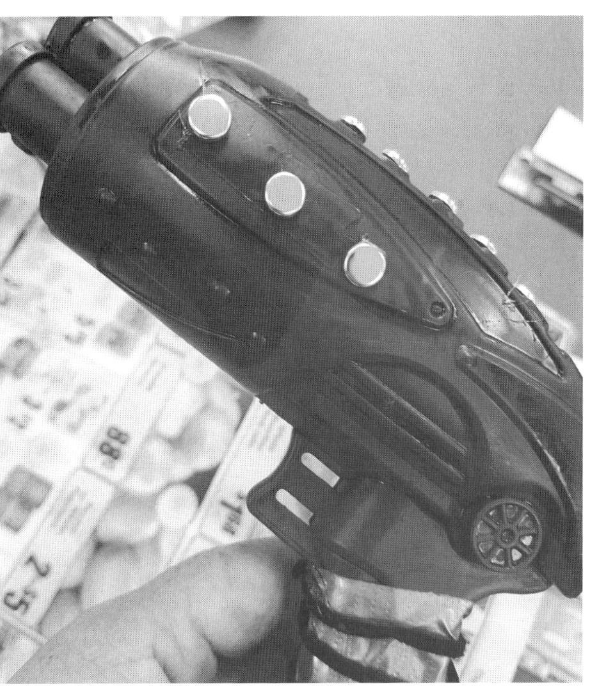

FIGURE 4.31

3. Apply the base coat of acrylic paint to the toy. Apply in thin coats using foam brushes or wide artist brushes. Let dry between coats, using a blow dryer set to a cool temperature and low speed if desired. Build up the base coat until the primer is covered (Figure 4.32).
4. Add layers of paint to create details and faux finishes. Use materials such as rags, sponges, and dry brushes to add texture and interest. Play with the paint until you get the look you like.
5. Distress your blaster, if desired, using sandpaper, dirt, and nail files to "rough up" your toy.
6. Protect your paint job with a clear acrylic spray sealant (Figure 4.33).
7. Add final decorative elements such as metallic elements, rivets, rhinestones, feathers, chains, etc.

Take It Further

Although they were written for the Steampunk aesthetic, Tom Willeford's two books provide many projects, techniques, and ideas written for beginners that are sure to inspire your young prop makers. Shawn Thorsson's book provides great tips for more futuristic styles. Superhero fans and future cosplayers may want to explore Svetlana Quindt's book for advanced techniques. Check your local library for:

- *Steampunk Gear, Gadgets, and Gizmos: A Maker's Guide to Creating Modern Artifacts*, by Thomas Willeford (McGraw-Hill Education TAB, 2011)
- *The Steampunk Adventurer's Guide: Contraptions, Creations, and Curiosities Anyone Can Make*, by Thomas Willeford (McGraw-Hill Education TAB, 2013)
- *Make Props and Costume Armor: Create Realistic Science Fiction and Fantasy*

FIGURE 4.32

FIGURE 4.33

Weapons, Armor, and Accessories, by Shawn Thorsson (Maker Media, Inc., 2016)

- *The Costume Making Guide: Creating Armor and Props for Cosplay*, by Svetlana Quindt (IMPACT Books, 2016)
- *Foamsmith 2: How to Forge Foam Weapons*, by Bill Doran (Punished Props, 2016)

Functional Enhancements

To understand how to modify a Nerf blaster, you first need to understand how Nerf blasters work. There are two types: spring guns and pump guns. Spring guns are more common. In spring guns, cocking the gun pulls back on a plunger inside a long barrel inside the body of the gun, compressing a spring at the back. The dart is placed in the outside end of the barrel. The trigger releases the spring, pushing the plunger forward and creating a puff of compressed air that launches the dart. In pump guns, the pumping action builds up air in a pressure chamber with an overpressure-release valve for safety. The trigger in a pump gun is attached to a metal rod that releases the air in the pressure chamber, sending it down the barrel and launching the dart. Each type of blaster has its advantages and disadvantages.

One of the most popular modifications to a Nerf gun involves replacing the barrel. Many Nerfers like to use their own homemade darts, called *stefans*, which have better range and performance. Replacing the barrel allows a Nerfer to use stefans as ammo. In addition, you can make the barrel longer, increasing accuracy, and tighter fitting, allowing for stronger launches. The most common barrel replacement material is PVC, which is expensive and easy to find at most hardware stores. More adventurous Nerfers have used metals such as brass and aluminum as well.

Another popular modification is to replace and upgrade the spring. A stronger spring will release more quickly and powerfully, creating more pressure in the launch and more distance when the dart is shot. Replacement springs are widely available at hardware stores, making it an easy and inexpensive upgrade. However, you have to take care. Using too strong a spring can potentially crack and damage the plastic mechanisms.

Another popular modification is removal of the air restrictor valve, which limits the flow of air in the barrel and slows the dart. Removal of the overpressure-release valve on pump guns allows a user to add more air to the pressure chamber, increasing the force of launch but also increasing the chance that too much air will be forced into the chamber and will rupture the chamber under pressure, causing injury. While both of these modifications certainly can increase the action of the gun, you may not want to allow them at Maker Camp for safety reasons. Either way, campers should wear safety goggles at all times when they use their blasters.

Making functional modifications to Nerf guns is a great way to explore mechanics and build skills with simple tools. However, you do want to take care that campers understand how to safely use items such as pipe cutters and/or a hacksaw before beginning. Depending on your budget or time constraints, you can perform some or all of the modification outline below or try your own upgrades.

Please note that every gun is a little different, so you will need to adjust your approach based on the model you choose. Modifications for many popular Nerf blasters are detailed online, so you should absolutely research your blaster first. Not all modifications will work on each model. If you are using off-brand blasters to spare your budget, you will likely be in uncharted territory.

As you go through any functional modification, be sure to take pictures on your smartphone frequently, especially if you open up the blaster. You may not easily remember how the pieces go back together. Taking a picture can be the difference between a functional blaster and one that no longer works.

Modification 1: Making Stefans

Homemade stefan darts will outperform any store-bought dart. They are inexpensive and easy to make. You may need to modify the barrel of your Nerf gun to use them, however.

Keep in mind that getting hit with a stefan may hurt slightly more than with a regular Nerf dart because it has no rubber cap and may be moving faster as the result of "backloading," which allows the dart to be closer to the source of the air pressure and the internal end of the barrel. Make sure that campers wear safety gear at all times. Do not allow close-range shots.

To make stefans, you will need to pick up a ½-inch backer rod. This is a cylindrical type of foam used to fill cracks and gaps between building materials. It may also be referred to as *precaulking rope* or *caulk saver*. It can be found at most hardware stores or purchased online. One hundred feet will cost under $15. You can make between 400 and 800 stefans depending on the length of the dart you choose to cut.

The dart will need a small weight at the tip. Number 6 metal washers (⅜ inch) are common. They'll cost about 5 cents each. BBs or fishing weights (such as small splitshots) are also common weights and may be even less expensive.

To use stefans, you will likely need to remove the air restrictor of your Nerf weapon, which can block the backloaded position that enhances the launch of a stefan. The air restrictor is usually a small piece with three prongs found with a spring. It can be completely removed. Also look for the barrel post, the long rod that is located in the stock. Nerf darts slip over it when it is placed in a Nerf gun. You'll want to remove the post.

Materials

- ½-inch foam backer rod
- A blow dryer
- Number 6 washers (⅜ inch outside diameter) or other weights
- Scissors or a craft knife
- Ruler or measuring tape
- Hot glue
- Safety goggles and gloves
- A pillow case (optional)

Process

1. Using the scissors or a craft knife, cut your backer rod. Standard Nerf darts are 2.75 inches long. Some Nerfers prefer shorter darts, as short as 1.5 inches. Experiment with what works best.
2. Using a blow dryer, heat the backer rod to straighten it. This will take 1 to 2 minutes of heat. Some Nerfers prefer to put the stefans in a pillow case and aim the hair dryer in, shaking the darts. This allows them to straighten many darts at once.
3. Use hot glue to attach the washer to one end of the dart. This weight will help your dart fly straight. One the weight is attached, use the glue to make a small dome over the weight to make the dart more aerodynamic (Figure 4.34).

FIGURE 4.34

Modification 2: Replacing the Spring

There are a whole host of popular spring modifications. Some are simple, like harvesting the spring from a similar Nerf gun and "double loading" a new gun, forcing two springs into the space meant for one. Similarly, some Nerfers add small objects such as coins or cardboard behind the spring to compress it. In both cases, the goal is the same: the more compressed the spring, the more tension it carries. When released, that potential energy becomes kinetic, forcing the spring forward more quickly and powerfully. This, in turn, created a stronger blast of air, launching the dart faster and harder.

Another option is to simply replace the spring. All springs have a *spring index*, with relates the diameter of the spring itself to the diameter of the wire in the spring. This ratio affects the strength of the spring by affecting the number of coils as well as the compression, extension, and torsion of the spring. In general, for compression springs like the ones used in Nerf guns, increasing the wire diameter a bit can increase compression, yielding a more forceful launch. Of course, this added force can damage your gun, so some Nerf experts recommend using a shorter spring over all to reduce the stress on the plastic pieces. For those who are interested in the somewhat involved engineering and mathematics behind spring modification, NerfHaven, an online website devoted to Nerf modification (nerfhaven.com), has excellent information on its community boards.

Materials

- A spring-powered Nerf-style blaster
- A set of screwdrivers
- A metal cutter, bolt cutter, or rotary cutting tool (such as a Dremel)
- A replacement spring of a thicker gauge than the original spring
- Safety googles and gloves

Process

1. Use the screw drivers to open the Nerf gun (Figure 4.35).

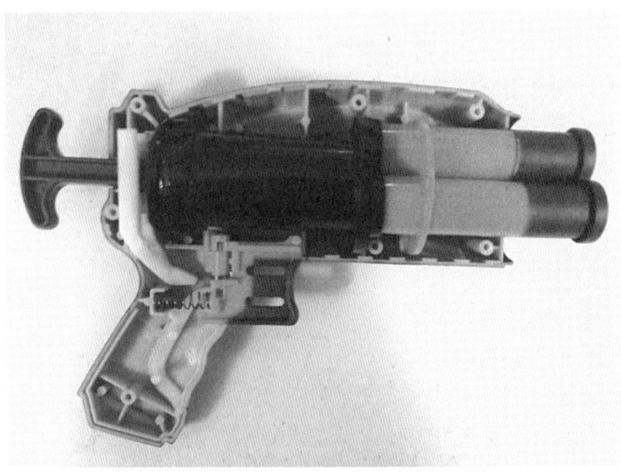

FIGURE 4.35

2. Locate the spring. It should be on a plunger near the back of the gun.
3. Remove the spring. The plunger should twist out or unscrew, allowing you to remove the spring (Figure 4.36).

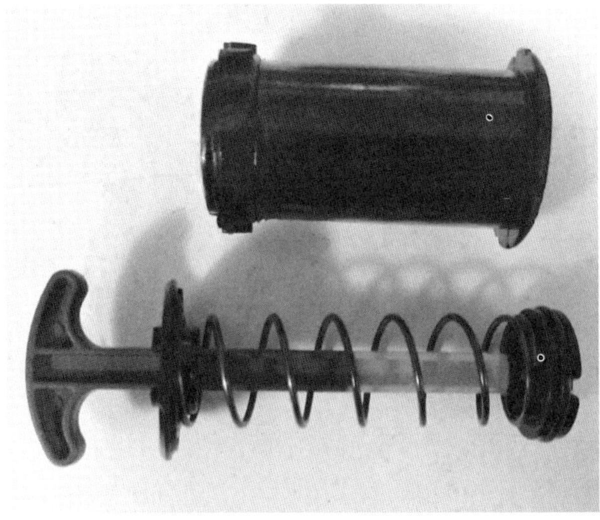

FIGURE 4.36

4. If needed, use the metal cutter or rotary tool to cut the spring down to the proper size (wear safety gear.)
5. Replace the spring. If it doesn't sit flush against the back of the plunger, add a few pennies or a piece of cardboard.
6. Screw the plunger back into place. Put the gun back together.

Take It Further

When you have the plunger out, you may want to replace the O-ring. The tighter the fit, the better is the seal between the plunger and the barrel, which means more of the pressure from the force of the air will be transferred to the dart, giving it more distance. You can add a tighter-fitting O-ring, a second O-ring, or use electrical tape to make the fit tighter.

Modification 3: Replacing the Barrel

Replacing the barrel is the most challenging modification and arguably the most effective for certain types of Nerf guns. If you plan to use stefans and replace the spring, changing the barrel to one that is sized for your new ammo and strong enough to withstand the stress of the new spring is a good idea. Replacing the barrel is considered an essential modification by many Nerfers.

The simplest replacement is a *single*, where the stock barrel is replaced with a stronger material. Of course, once you master this technique, you can create more complicated configurations, such as a double barrel or a coupled barrel with two sizes of materials that allows the darts to be backloaded quickly and easily. We'll be reviewing a basic barrel replacement. Further references for more complex techniques are available on YouTube, Instructables.com, NerfHaven, or Nerf Wiki.

One of the most common materials used is ½-inch chlorinated polyvinyl chloride (CPVC), which is a plastic tubing usually used for plumbing. Its diameter makes it possible to front load traditional Nerf darts while also backloading stefans if you want to make internal modifications such as removing the air restrictor. CPVC is inexpensive and can be found at most hardware stores. PVC will also work, although you may want a diameter of less than ½ inch. You want to choose a tubing that will fit your stefans without leaving a lot of room around them so that you don't lose any air pressure but isn't so tight that you greatly increase friction.

You'll need a good adhesive. You can use hot glue, which sets quickly but isn't as strong as other adhesives. Superglue works well, but you need to use sandpaper on the surfaces for it to hold. Two-part epoxy is super strong, but it isn't easy to work with and can take a long time to set. A Nerfer standard is Plumber's Goop, which is traditionally used to seal leaks and cracks. It takes a while to set, but it holds very well.

As for the length of the barrel, you'll likely want to mimic the length of the barrel on the gun you have, although obviously you can go longer or shorter. Too long a barrel can actually slow your launch. Try different lengths to see what works. You can temporarily hold the CPVC in place with electrical or duct tape.

In most cases, you'll be removing the barrel of the blaster from the outside. In some cases, you may be able to open the blaster and feed a replacement barrel into the toy. You will need to evaluate your particular blaster.

Materials

- A Nerf-style blaster
- ½-inch CPVC
- A hacksaw
- Clamps as needed
- A PVC pipe cutter (optional)
- Sandpaper
- Tape measure
- Permanent marker or pencil
- Your choice of adhesive (see above)
- Safety goggles and gloves

Process

1. Put on safety goggles and gloves. Clamp your gun to your worktable with the barrel hanging over the edge.
2. If needed, use the hacksaw to remove the barrel of the blaster near the front of the blaster. Leave some of the original barrel on, up to 1 inch. In some cases, most of the barrel is actually internal, and you will need to open the gun to remove it (Figure 4.37).
3. Use sandpaper to smooth the sawed barrel of the blaster.
4. Measure and mark your length of CPVC. Cut your CPVC using the hacksaw.

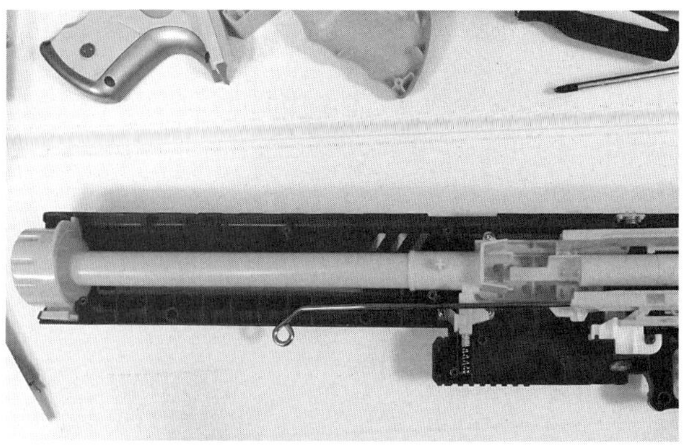

FIGURE 4.37

5. Slide the new barrel over the remaining tip of the previous barrel. Use your adhesive to attach the new barrel (Figure 4.38).
6. Let the adhesive set for the time required, according to the directions. Do not try to shoot your new Nerf gun before the new barrel has had time to set.

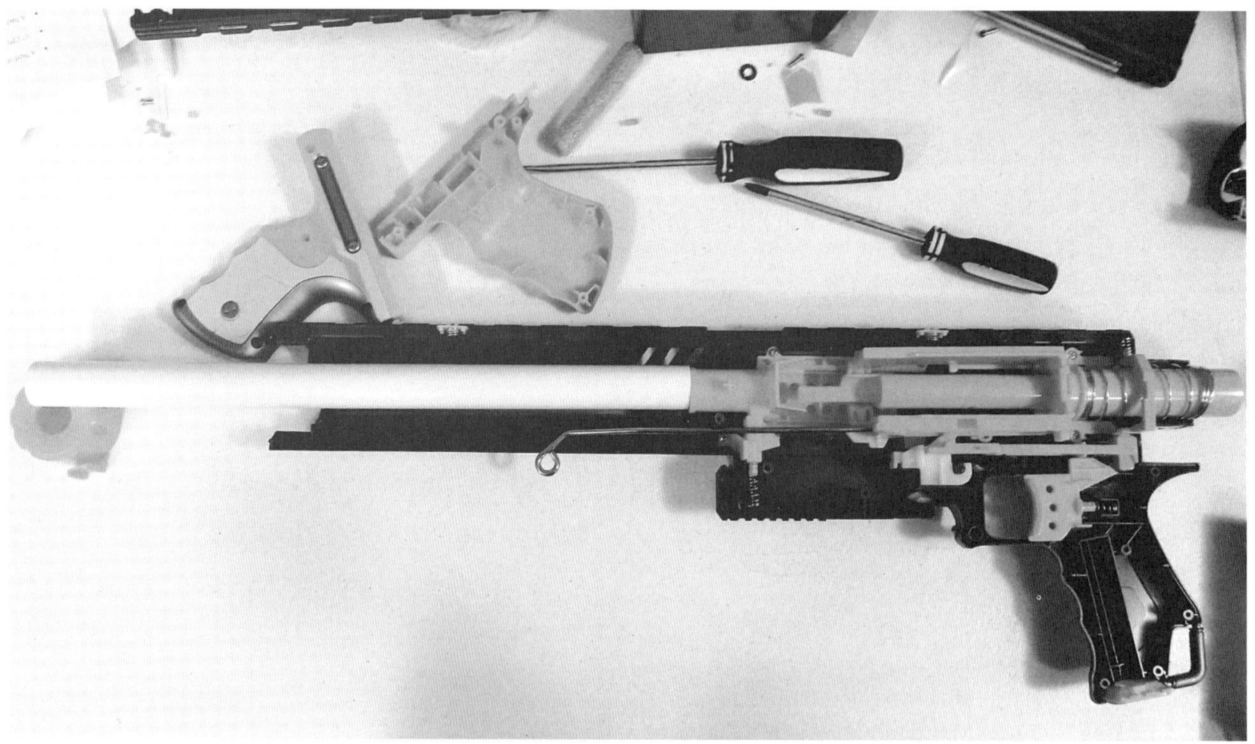

FIGURE 4.38

Creating Nerf Targets

Once your campers have modified their Nerf gun, they'll probably want to put them to use! Have your campers create simple targets at which to launch. These are all easy to construct.

Bull's-Eye

Paint a bull's-eye on the side of a cardboard box. This is easy and fast. Plus, it'll stand on its own.

Bean Bag–Style Target

Use a cardboard cutter or craft knife to cut different-sized circles into the face of a cardboard box. Paint the front.

Drink Cup Targets

Poke a hole in the base of a plastic drink cup. Thread string through the hole, tying a knot at one end inside the cup. If desired, tie a small bell inside the cup. Tie the cups to a broom handle, piece of PVC pipe, or scrap 2 × 4 so that they hang down from the top at various levels. Lay your targets across two chairs, stools, or a set of sawhorses.

Paper Plate Targets

In a large corrugated cardboard box, cut squares larger than your paper plates. Decorate the plates as you like. Tape or glue a drinking straw on the back of each plate. Poke a bamboo skewer into the folds of the corrugated cardboard on each side of a square, horizontal to the face of the box. Feed the skewer into the drinking straw on the plate so that it faces outward and spins when hit.

Pool Noodle Targets

Pull pool noodles into circles, and duct tape their ends together. Either hang the circles from trees or tape them together to create a matrix of circles. Mount the matrix with a vertical pool noodle on each side. Place a tent stake or a wooden dowel in the ground, and fit the holes in the two vertical pool noodles over them to hold the target upright.

Water Bottle Targets

Fill disposable water bottles with varying amounts of sand or water. Line them up along a fence, wall, or table edge. Alternatively, stack empty soda cans in pyramids or other shapes.

Creating an Obstacle Course

Creating an obstacle course for a Nerf battle is pretty easy to do. You may already have a good area near your camp, such as a playground or a park. If not, here are some easy items that you can use to create a course in any open field.

Make sure each team has a "home base" where team members can return to rest, reload, or otherwise regroup. Remind campers that they can fire at close range. Campers should all wear safety goggles during a Nerf battle.

Combine these obstacles with your DIY targets, and your campers will have ample opportunity to play:

- Bales of hay
- Large cardboard boxes
- Small tents
- Play tunnels
- Sawhorses draped with sheets or drop cloths
- Wooden pallets secured with bricks or cinder blocks

PROJECT 8
PVC Marshmallow Shooter

COST: $

MAKE TIME: 15–30 minutes

It's amazing what you can make out of PVC. It's inexpensive, easily available, sturdy, easy to cut, and very adaptable. This simple marshmallow air gun is one of my campers' favorite things to make. Each costs about $2 to make, and they can be customized in configuration or decoration.

Note that I don't use PVC cement to join the pieces. Not only are the chemicals flammable and toxic, but cementing the pieces together makes it hard to remove any marshmallows that get stuck inside.

Materials

- 2 feet of ½-inch internal diameter Schedule 40 PVC
- 2½-inch PVC 90-degree turns
- 2½-inch PVC caps
- 2½-inch PVC tees
- Hacksaw or PVC cutter
- Sandpaper
- Measuring tape
- Marker
- Spray paint or color duct tape
- Mini-marshmallows
- Goggles

Process

1. Use your measuring tape to mark pieces to the following lengths along your 2-foot section of PVC: 6 inches, 4 inches, 4 × 3 inches, 2 inches.
2. Using the hacksaw or PVC cutter, cut the pieces, starting with the smallest first.
3. Gently sand off any rough ends.
4. Put the pieces together as shown in the diagram (Figure 4.39).
5. If desired, spray paint or cover the launcher in duct tape to decorate.

To launch a marshmallow, put on your goggles, place a marshmallow in either end of the launcher, place your mouth against the 4-inch section, and give a strong puff of breath.

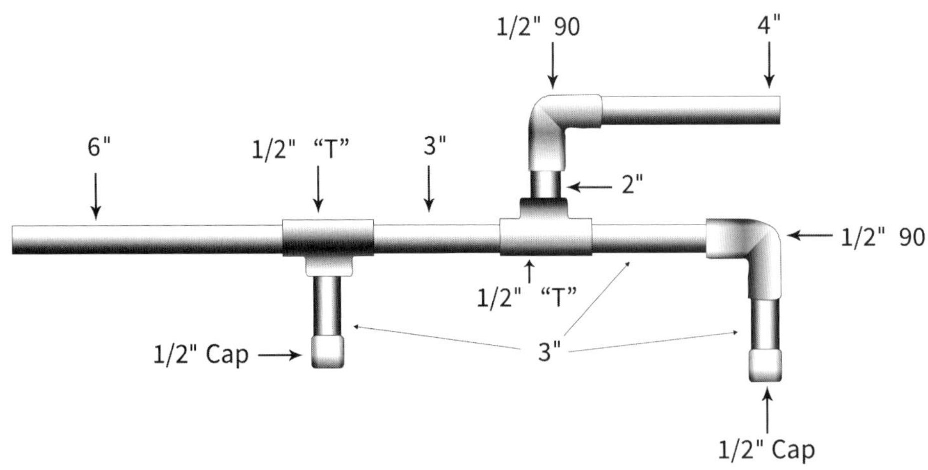

FIGURE 4.39 Diagram by Jennifer McBride.

PROJECT 9
Marshmallow Poppers

COST: Free–$

MAKE TIME: 5–15 minutes

When I was planning my superhero-themed Maker Camp for 2018, I wanted a simple activity for the end of the day that we could take outside and play with. This was perfect. It's easy to make, super cheap, demonstrates some cool physics, can be customized and decorated, and uses marshmallows as ammo, so any ammo left behind dissolves in the rain.

Because it was such a huge hit at Maker Camp, I brought it to the World Maker Faire in 2018. By the end of Saturday, we were out of pool noodles. Because it was fall, we couldn't find any in stock anywhere. Instead, we picked up inexpensive foam pipe insulation, which worked just as well.

Materials

- Pool noodle or foam pipe insulation
- Duct tape
- 12-inch latex balloon
- Mini-marshmallows
- Scissors
- Serrated knife such as a bread knife

Process

1. Cut your pool noodle or pipe insulation into 3- to 4-inch sections using the knife.
2. Tie a knot in the neck of the balloon as close to the body of the balloon as you can.
3. Cut the rounded top of the balloon off, ½ to 1 inch from the top.
4. Stretch the balloon over the pool noodle, pulling the knot as close to the base as possible, centering it over the hole.
5. Secure the balloon with duct tape.
6. Decorate as desired (Figure 4.40).

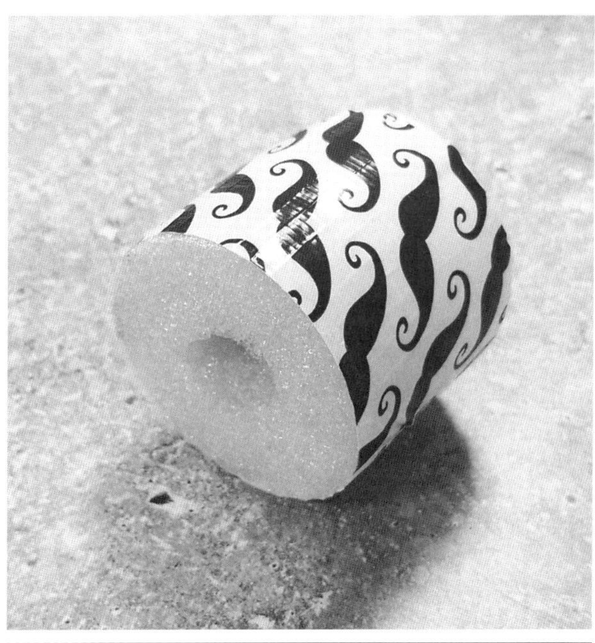

FIGURE 4.40

It can help to put a small piece of foam in the hole at the back of the noodle, pressure fitted near the balloon. Not only does this keep the marshmallow from falling into the balloon, but it also seems to help concentrate the force of the balloon snap, giving the marshmallow a better launch.

To launch a marshmallow, put on goggles, place a marshmallow in the hole, pull back on the knotted neck of the balloon, and then release.

PROJECT 10
Creating Resin Action Figures

COST: $$

MAKE TIME: 30 minutes, plus time for drying and curin

Casting and molding are traditional hand crafts that provide a lot of opportunity for campers to explore new materials and learn chemistry. This 6,000-year-old technology involves creating a solid mold and then pouring a liquid into it that hardens to create a new item. Using this process, you can make anything from plastic toys to chocolates.

Of course, you can make a mold out of many materials. For example, you can 3D-print a mold or use a vacuum form to create one. Traditional methods use glass, clay, ceramics, stone, wood, plastic, and metal. There are several ways to create molds, but many processes involve creating a permanent, reusable mold that can be used for many casts. Usually, a release agent is used to make sure that the new object doesn't stick in the mold.

For this project, we'll be making simple molds out of common materials and then casting with a two-part craft resin. Campers can use small plastic figures, such as Lego minifigs, to create the molds. This figure is called the *positive*. A mold is a negative impression of the original formed by contact with the original. A cast is a positive impression of the original formed by contact with the mold.

The mold for this project is made using 100 percent silicone caulk. You can't substitute other products because they won't harden correctly. Because this caulk does release a lot of fumes, you're going to want to work in a well-ventilated room.

Unfortunately, you can't use silicone alone to make a mold. Silicon cures as a result of the water in the air. It is usually used in thin applications, like caulking. However, when making a mold, you need much thicker sections. The water can't get in to cause the reaction that allows the mold to cure, so you end up with a mold the never hardens.

Acrylic paint is an emulsion of water, acrylic polymer, and pigment. By including it in our mold mixture, we mix in water, which helps the silicone to cure consistently throughout. Because acrylic is also a binder, it helps the mold to cure more quickly and makes it sturdier. A few drops of baby or mineral oil help to create a more flexible, soft mold, which makes it easier to remove casts later.

Corn starch is used to make the silicone less sticky to work with. The starch also traps moisture from the air, which gets mixed in the silicone and helps it cure. Some people prefer baking soda or liquid glycerin instead, so you can experiment if you like.

There are also commercially available silicone mold materials such as Smooth On, as well as skin-safe alginates and melt-and-pour molding materials such as ComposiMold that you can buy. These may be preferable if you are working with younger campers.

When it comes to what you plan to make your casts out of, you have a lot of options. You can use a lightweight air-dry clay, such as Model Magic, or polymer clay. Plaster of paris will also work, as long as you are gentle with the unmolding. Some people have even filled the molds with hot glue to create casts.

I suggest a two-part epoxy casting resin. This kind of resin includes a liquid plastic and a separate hardener. When mixed, a chemical reaction occurs, creating a hard, clear polymer. You can add dyes or glitters to the mixture before hardening or use paint afterwards. It's also possible to embed small sequins, beads, or similar items in the resin.

Materials

- Cornstarch
- Silicone I all purpose 100 percent caulk
- Caulk gun
- Baby oil or mineral oil
- Inexpensive craft acrylic paints
- Disposable bowl or cup
- Popsicle sticks
- Objects to cast
- Vaseline
- Vegetable oil spray or a commercial release agent such as Ease Release 200
- Two-part plastic resin
- Small disposable measuring cups
- (Optional) Fine-grit sandpaper, paint brushes, acrylic paints

FIGURE 4.41

FIGURE 4.42

Process

1. Either by estimating or measuring, determine the amount of material you'll need to make a mold large enough for your item to be pressed into with space around it and under it to make a solid mold. Assume that you will need ¼ to ½ inch surrounding your item.
2. You'll need roughly a 2:1 ratio of silicone to cornstarch. Note that you will need to work quickly because the mixture becomes too stiff to work with after 10 to 15 minutes.
3. Place the silicone tube into a caulk gun, and squeeze the silicone into a bowl.
4. Add up to 1 teaspoon of acrylic paint and several drops of oil. Use a Popsicle stick to mix the paint into the silicone (Figures 4.41 and 4.42).

5. Add the cornstarch, and mix it in. Once it is basically mixed, use your hands to knead the mixture as needed until a uniform putty is formed (Figure 4.43).

FIGURE 4.43

6. Roll the putty into a ball, and press it slightly to flatten. Be sure that there are no bubbles (Figure 4.44).

FIGURE 4.44

7. Lightly dust your item with cornstarch or lightly coat it with Vaseline.
8. Press your item into the mold. Take care not to press the item in so far that pieces become embedded in a way that makes them tough to remove. Gently pull the item out (Figure 4.45).

FIGURE 4.45

9. Allow the mold to cure for 20 to 30 minutes (Figure 4.46).

FIGURE 4.46

10. Spray the mold lightly with vegetable oil or release.
11. Prepare your resin by mixing together equal amounts of each part. Follow the instructions for your specific resin. If desired, mix in glitter or dye.
12. Pour the resin into the mold, and gently tap the mold to release bubbles. Allow the resin to set according to the directions given by the manufacturer (Figure 4.47).

FIGURE 4.47

13. When the cast has set, gently stretch the mold a bit, and push the cast out of the mold (Figure 4.48).

FIGURE 4.48

14. If desired, sand the resin object lightly. Then paint the item.

PROJECT 11
Nerdy Derby

COST: Free $$

MAKE TIME: 45–60 minutes

If you've ever been to the World Maker Faire in New York City, you may have had the opportunity to experience a Nerdy Derby. Loosely based on the Boy Scout tradition of the Pinewood Derby, a Nerdy Derby is an all-out, no rules, self-propelled minicar race. The official race has a wooden curved track and formal races, but even without the complex track, it's easy to capture the DIY fun of a Nerdy Derby at camp.

If you have the skills, money, and time, you can, of course, build a complex track out of wood, as described on the official Nerdy Derby website (www.nerdyderby.com), but if time, budget, and storage space are challenges, there are other ways to create a track. For example, an old shelf or a plank of wood with one end rested on a table and the other on the floor will work. I personally love using a 10-foot piece of vinyl gutter material, available for under $5 a piece at your local home-improvement store. The sides make it easy for cars to stay on the track, it is long, and it is lightweight (Figure 4.49).

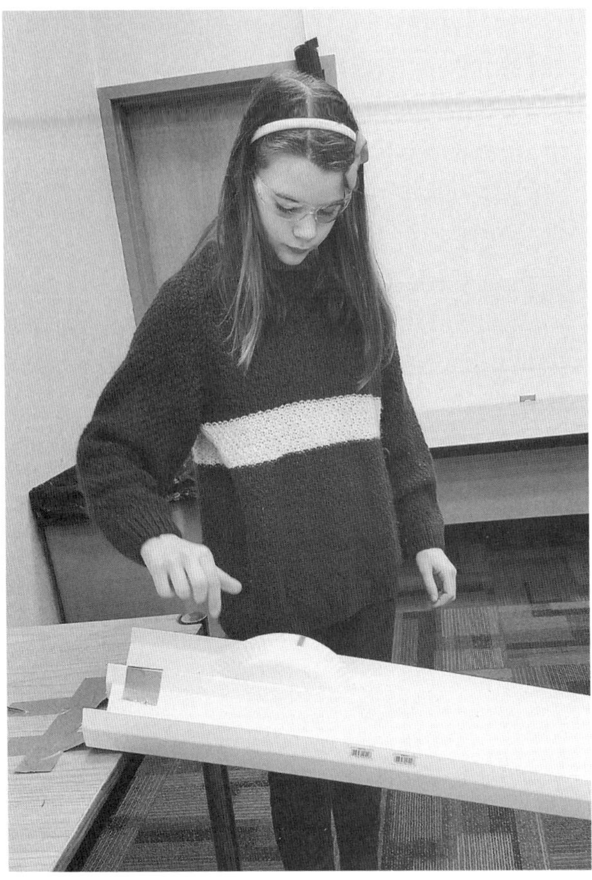

FIGURE 4.49

I usually set a few rules for the cars. First, a car is defined as a unit having a body, with at least two wheels, and an axle to connect the wheels. Second, either the wheel or axle must be free moving. Lastly, the car and all the wheels must fit on the track. These rules eliminate the temptation to just roll a marble down the track and call it a car.

As far as materials go, get creative! Car bodies can be crafted out of cardboard, toilet paper tubes, upcycled Styrofoam trays, disposable water bottles, plastic or paper cups, rolled paper plates, binder clips, clothespins, and pretty much any other solid material. For wheels, try metal washers, bottle caps (plastic and metal work), old CDs, repurposed wheels from old toys, wooden hobby wheels, wine corks, wooden thread spools, and even wagon wheel pasta. The axle will depend on the wheels selected, but bamboo skewers, pencils or pens, small dowels, flattened paper clips, drinking straws, nuts and bolts, nails, pipe cleaners, and brass paper fasteners are all great choices. You'll also want to include items such as old batteries or pennies to be used as ballast.

You'll need a variety of adhesives such as hot glue, tacky glue, superglue, duct tape, masking tape, and invisible tape, as well as wire and string. Also, you'll need tools such as scissors, X-ACTO knives, box cutters, cardboard cutters, awls, hole punchers, wire cutters, diagonal cutters, pliers, and needle-nose pliers. You may also want to include craft supplies such as stickers, markers, paint, feathers, and rhinestones. Of course, you don't want to forget safety equipment such as gloves and goggles (Figure 4.50).

FIGURE 4.50

Encourage your campers to prototype, rework, and improve their initial designs. It can be a good idea to set up test tracks for campers to test their cars on. Common problems are uneven or skewed axles, wheels that don't hit the track quite right, cars that are too wide for the track, axles or wheels that don't spin cleanly or freely (usually the result of an ill-fitting axle or too much adhesive), and weight placed in the wrong location of the car. It can take several dozen runs to get things just right.

Once everyone is happy with his or her design—or when the designated build time is over—it's time to race. You can either use a stopwatch for timed trials or just compete head to head in elimination challenges. If you like, you can offer prizes or ribbons for the faster, more creative, or most beautiful car, although often bragging rights are enough.

Materials

- Several 10-foot pieces of vinyl gutter
- Recycled/upcycled and craft materials for the car body, wheels, and axles
- Glue: hot, tacky, super, etc.
- Tape: duct, masking, invisible, etc.
- String and wire
- Scissors
- Cutters of various types: craft, box, carboard, wire
- Hole punches and awls
- Gloves
- Goggles

Process

1. Set up the race track and practice track. Use duct tape to secure the tracks to the table and floor as needed.
2. Set up multiple workstations with materials such as scissors and tape. Put out safety gear at each station.
3. Set up a hot glue station with volunteers to help younger campers.
4. Set up a station with materials for building the cars.
5. Review the rules of the competition with all campers.

6. Set a timer and begin building.
7. When complete, race the cars. Cheer for everyone.
8. When the races are over, take time to debrief about what worked and what didn't (Figures 4.51, 4.52, and 4.53).

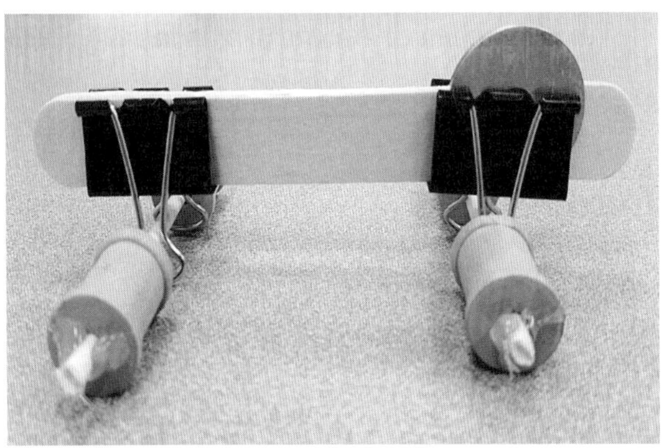
FIGURE 4.51

FIGURE 4.52

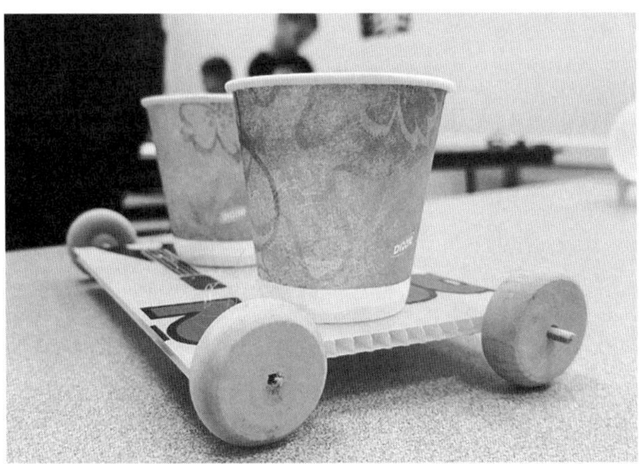
FIGURE 4.53

CHAPTER 5

The Campfire

No camp is complete without some time spent with friends around a campfire. Roasting marshmallows, singing together, and telling spooky stories are tried-and-true camp activities. Whether you have access to an outdoor fire pit or need to build your own battery-powered version, take time to gather around your campfire.

PROJECT 1 Color-Changing Flames

COST: $

MAKE TIME: 15 minutes

Color-changing flames may just be my favorite thing to demonstrate in science class. Usually I use an open flame from a Bunsen burner or propane torch and spray metallic salt solutions into the flame. It's incredibly engaging and exciting. On top of all that, color-changing flames demonstrate some really neat chemical reactions and principles of astronomy. Luckily, it's pretty easy to take that science lesson from the classroom into camp.

Instead of liquid solutions, we can make color-changing discs out of common materials. The basic egg carton and wax fire starter are common camping items in most circles, but here we take a similar idea and bump it up a notch.

You can order a variety of metallic salts from science supply companies. Each will give your different flame colors:

Blue	Cupric chloride
Red	Lithium chloride
Red	Strontium chloride
Orange	Calcium chloride
Purple	Potassium chloride
Yellow	Sodium chloride

Of course, there are other options if you do not want to or cannot order from a science supply house. Calcium chloride is routinely sold to melt ice in winter. Potassium chloride is sold as NuSalt in grocery stores and as a water softener alternative in hardware stores. Sodium chloride is just table salt. However, your best off-the-shelf pick is copper sulfate. Sold in hardware stores as a root killer, it makes an intense blue-green flame.

In all cases, a larger crystal is desirable to produce the most noticeable color change. You'll also want to minimize the amount of wax, using only enough to hold your creation together. Don't mix chemicals within each disk because that can muddy the color change.

When making these disks, campers must wear gloves and goggles. You should burn your disks outdoors only, just to ensure that there are no respiratory reactions. Remind campers that they are working with chemicals that can make them sick. Do not do this activity with young campers.

As to the science, when a metal or metal salt is burned, the input of heat as thermal energy excites the electrons in the metal atom to a higher energy state. The electrons can't stay at that energy level for long, so they release the added energy in the form of light until they reach stability. Each metal atom releases a specific frequency of light when it is burned, resulting in different colors.

Materials

- Your choice of solid chemical
- Paraffin wax or leftover candles
- Dryer lint or wood shavings
- Paper egg cartons (or paper muffin cups)
- A double boiler

Process

1. In the double boiler, melt the wax. Remove any wicks, if necessary. You'll need less than a quarter cup of wax per disk.
2. In each egg carton well or muffin cup, add a little dryer lint or a wood chip. Then add up to a tablespoon of chemical. Top this off with a little lint or a woodchip (Figure 5.1).

FIGURE 5.1

3. Pour the melted wax into the well or muffin cup, adding just enough to hold things together (Figure 5.2).

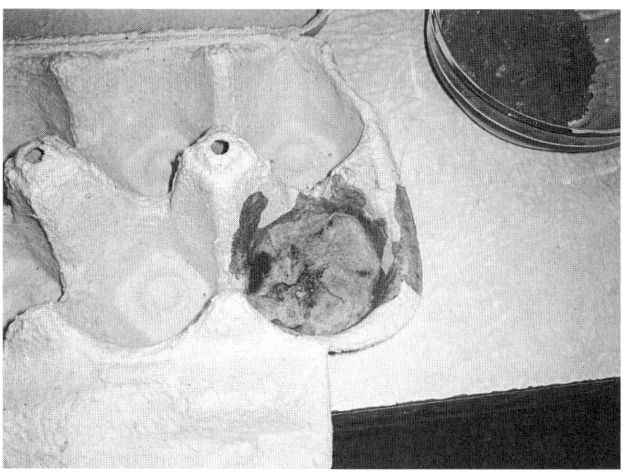

FIGURE 5.2

4. Allow the wells or cups to cool and harden completely. Rip apart the egg carton or remove the muffin cup to make individual disks Store in plastic zip-top bags or containers until use.
5. After building a fire and creating a good bed of coals, carefully toss the disks one at a time onto the fire and enjoy the show (Figure 5.3). This works best if you use several at a time after darkness has fallen.

FIGURE 5.3

PROJECT 2
Roasting Spy Marshmallows

COST: $

MAKE TIME: 15 minutes

You can't have camp without roasting a few marshmallows, but Maker Camp marshmallows should be better than the rest. An easy way to make roasting marshmallows even more fun is to make spy marshmallows. The heat of toasting them will reveal a secret message. It's easy to do and demonstrates a few cool scientific ideas.

Marshmallows are mostly sugar and gelatin. Those two components are the reason for their roasting ability, which gives them toasty, sweet flavors and beautiful golden hues. First, the heat melts the gelatin in the marshmallows. Then the amino acids react with the sugar through a complex set of chemical processes called the *Maillard reaction*. This is the same process that browns meat, adding flavor and color. As the temperature rises, the sugars start to melt and break down, creating aromatic compounds that yield amazing buttery, toasted flavors.

Both of these reactions are greatly affected by pH—how acidic or basic a material is. This is the key to this project. An increase or decrease in pH causes the marshmallow to brown more quickly. So, if we create areas that have different pH levels, we can make some spots brown first while other areas remain white.

As for roasting, even if you don't have a campfire available, a hotplate or stove provides enough heat to get your marshmallows toasty. It goes without saying that precautions need to be taken, including working on heat-safe surfaces and having a fire extinguisher nearby.

Once the message is reveled, campers can certainly eat the evidence. After roasting, the marshmallows are perfectly safe to eat, although they may taste a bit salty.

Materials

- Marshmallows
- Bamboo skewers or marshmallow skewers
- A campfire, stove, or hotplate
- Baking soda
- Water
- Small disposable cups, 3 ounce
- Popsicle sticks
- Cotton swabs or paint brushes

Process

1. Add a teaspoon of baking soda to a cup, and fill the cup about two-thirds of the way with warm water. Mix well.
2. With a cotton swab or paint brush, draw or write on a marshmallow with the baking soda solution.
3. Let the marshmallow dry completely.
4. Place the marshmallow on a skewer. Hold it over your heat source. The area treated with the baking soda will brown first, revealing the message (Figure 5.4).

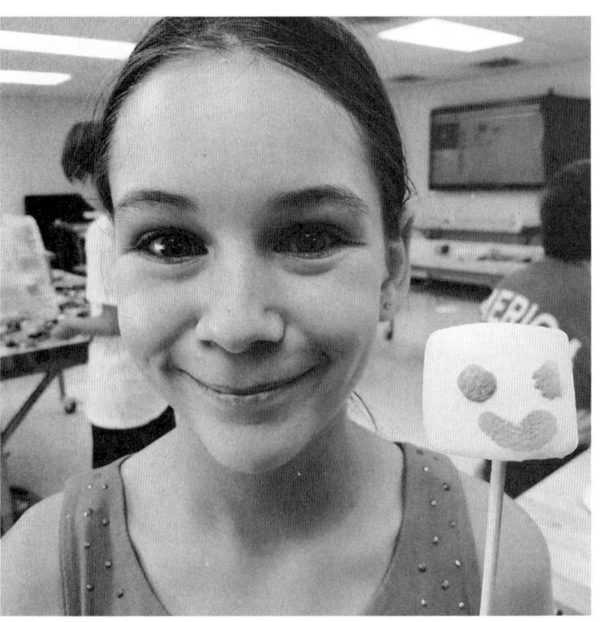

FIGURE 5.4

Take It Further

Once campers observe the original reaction, try different solutions to use for hiding your messages. Try lemon juice, milk, vinegar, and other acidic or basic liquids. What creates the best result?

PROJECT 3
Building a LED Campfire

Many Maker Camps are held during the day at places such as schools and libraries, so you probably don't have access to a fire pit around which to sing camp songs. But that's not a problem! We're makers, and we're adaptable, creative people.

Below are three different ways to create a faux campfire for your Maker Camp. The first is meant as a large decorative display. The second is an inexpensive tea-light version that each camper can build. The last uses a Circuit Playground Express, a small direct-current motor, and craft supplies to create a realistic flame effect.

Decorative Camp Fire Display

COST: $

MAKE TIME: 15–30 minutes

This is meant as a larger display to be used by all the campers.

Materials

- Several larger stones to create a fire circle
- Firewood or logs
- Green or brown round tablecloth (optional)
- Tissue paper in red, orange, and yellow
- Several stands of battery-powered LED lights in white or warm white

Process

1. Lay down the tablecloth.
2. Place the stones in a circle around the tablecloth.
3. Stack the wood in the center of the circle. Two easy ways to stack the wood are the "teepee" formation and the "log cabin" formation. The teepee structure is where you lay the wood against each other to form a cone-like frame. For the log cabin structure, lay two logs side by side, with several inches in between. Stack the next level with two more logs on top of the first two, side by side, but in the opposite direction, making a hashtag configuration.
4. Tuck a string of lights under the stacked wood.
5. Fold and crumble the tissue paper to create paper flames. Tuck the tissue paper in among the wood.
6. Tuck additional strings of lights into the paper flames.
7. Turn on the lights and enjoy.

Individual Campfires with Tea Lights

COST: $

MAKE TIME: 15 minutes

This is an inexpensive craft that you can make for under a $1 per camper. To gather around the campfire as a group, just bring all the mini campfires together on the floor or on a table and light them up together.

Materials

- An upcycled CD
- Paint and paint brushes or construction paper
- Small rocks
- Hot glue or white glue
- Glue stick

- Squares of tissue paper in red, orange, and yellow, about 6 × 6 inches
- A flameless LED tea light with a flickering "flame"
- Craft knife

Process

1. Paint the back of the CD with green or brown paint. If you have limited time, trace the CD on brown or green construction paper. Use the glue stick to attach the paper to the CD.
2. Use glue to attach small rocks around the outside of the CD. Let dry.
3. Layer the squares of tissue paper, rotating each layer so that they are each at different angles. Use the glue stick to attach each square together in the center.
4. Place the tea light in the center of the paper. Gently pull the paper up and over the tea light; gather the paper and twist it a bit so that it stays above the tea light, mostly covering it. Use a craft knife to expose the switch on the bottom of the tea light.
5. Glue the tea light and tissue paper combination to the center of the CD and allow to dry. Make sure that the switch on the tealight is centered over the hole in the CD so that the "campfire" can be turned on and off.
6. Turn the tea light on and enjoy.

Individual Campfires with Circuit Playground

COST: $$

MAKE TIME: 15 minutes for coding, 30 minutes to build

If you're looking for a simple coding and electronics project, this is it. It uses Circuit Playground Express (CPX), which is a great little board produced by Adafruit Industries. For $25, it packs a lot of useful electronics into a small, easy-to-use space. The CPX includes a ring of color-changing neopixels that you can control through code. It also has several sensors, a speaker, and much more onboard. We're using some of the simple features here, but a more complex project using the CPX is outlined in Chapter 3.

To hold the electronics, I used the ubiquitous red party cup. Other cups will work, of course; just make sure that they are large enough to hold the CPX with additional space for cables. Any color will work. If desired, paint and decorate your cup before fitting the electronics into it.

For the medicine cup, I upcycled one from a bottle of cough syrup, but you can buy them at a pharmacy or a big-box store. The base of the cup must be larger than the base of your motor, and the outer rim must be smaller than the CPX. It must be clear or translucent to allow light to pass.

The most challenging part to find is the motor. You can find them at hobby stores, order them online from electronics distributers, or even hack an electric toothbrush or toy. Just make sure that your motor is powered by 3 volts; otherwise, this project won't work.

You will also need a small propeller that fits the shaft of the motor. You can usually find one of these in the same place you found the motor. Just be sure that it will not be wider than the diameter of the cup you are using. You can make your own from a bit of scrap plastic, such as the top from a take-out container. Playing with blade design is a great engineering activity. Just make sure that the propeller attaches firmly to the motor shaft.

When selecting a battery pack, make sure that it has wires that are long enough to give you room to work. The cable should be at least 6 inches long. The switch is absolutely necessary because otherwise your motor will run nonstop. This switch will turn your campfire on and off.

Materials

- Circuit Playground Express
- 3 × AAA battery holder with on/off switch and JST 2-pin connector cable
- Direct current toy/hobby motor, 130 size, 3 volt
- Propeller
- 3 AAA batteries
- 2 alligator clip test leads
- 2 plastic 18-ounce party cups
- 1 clear medicine cup
- An awl or screwdriver
- Scissors and/or craft knife
- Hole punch (optional)
- Permanent marker
- Tape measure
- Colored tissue paper
- 2 drinking straws
- Paint and paintbrush (optional)
- Small stones and twigs
- Hot glue or white glue (optional)
- Invisible tape
- Small double-sided adhesive foam squares (optional)
- Safety goggles and gloves

Process: Coding the Circuit Playground

This code is based on the project "Circuit Playground Jack-o'-Lantern," by Phillip Burgess for Adafruit. For a full discussion of the math behind this flickering flame, see https://learn.adafruit.com/circuit-playground-jack-o-lantern/halloween.

Like Scratch, MakeCode is a visual language that uses interlocking blocks of code to create computer programs. Each group of blocks is color-coded by purpose. The window is divided into three sections. On the left, there is an interactive model of the CPX board that will virtually follow your code, a central panel holds the code blocks, and on the right is the Scripts area where you will arrange your code. Note that you can also select "JavaScript" with the top slider if you prefer that coding environment.

1. Go to https://makecode.adafruit.com/. Click on "New Project."
2. Click on the green "Loops" menu. Drag the "on start" block to your Scripts area. This loop will initialize the CPX.
    a. Click on the "Light" panel, and drag the "set brightness" block into the Scripts area and place it inside the "on start" loop. In the white bubble, type "255." This will set the brightness to the highest level.
    b. Click on the "Variables" panel. We will create the variable "lvl," which will control the brightness of the lights; a "dir" variable, which will help us control how much the light brightness changes; and the "gamma" variable,

which will control the color of the light. Using these variables, we'll create a semi-random–looking flame. To create the variables, click on "Make a variable," and enter the appropriate name.

c. Select the "set [variable] to [value]" block. Drag this into the Scripts area, and place it into the "on start" block. There is a small downward arrow that marks a drop-down menu. Click on the drop-down menu, and select "lvl." In the white bubble, enter "128."

d. Repeat step 2c. Set the drop-down menu to "dir." In the white bubble, enter "5."

3. Use the same drag and drop to re-create the "Forever" loop. Take care to nest the "if else" and "if" loops correctly. If you prefer, you can download the code file from the website referred to in the introduction. Here's a description of what's happening.

   a. The "Forever" loop will keep running the code within it for the entire time the CPX is powered.

   b. The first "if" loop ensures that the brightness does not go over its maximum level of 255. If the "lvl" variable goes over 255, it will reduce that by 5.

   c. The entire next "else" loop is used to create a semi-random flickering effect.

   d. The first "if" loop checks to see if the brightness has gotten too low. Although it can go to 0, we don't want to black out our fire, so if the level reaches 16 or lower, the code will bump it up by 5.

   e. The next nested "if" loop is the randomizer. The "pick random" block selects a random number between 0 and 10. If the result is 0, it multiplies whatever the "dir" variable is at present by –1. Essentially, this allows the code to roll a die and reverse the level of the brightness by a bit, creating a flicker that happens every now and again.

   f. After the "if else" loop, we set the "gamma" level, which varies the color.

   g. For any one of the neopixels on the ring of the CPX, you can set the value for red, green, or blue (i.e., RGB) to anywhere between 0 and 255. A value of 0 would mean none of a color, whereas 255 would be 100 percent of that color. That means 0, 0, 0 would be black, and 255, 255, 255 would be white. Meanwhile, 255, 0, 0 would be red; 0, 255, 0 would be green; and 0, 0, 255 would be blue. This system allows for 16,777,216 colors!

   h. The "gamma" variable works by taking the variable "lvl," which we already have, and multiplying it by itself, so it is never negative, which wouldn't work for an RGB number, and then divides it by 255, so it stays within the necessary 0–255 range. The result is that the value is always between 1 and 255.

   i. The blue "set all pixels" block sets the color on the neopixels in the ring. The value you set as "gamma" will be your dominant color. In our case, we've set it to red. Then we have green set to "gamma" divided by 8, so it is always less than red. Blue is even less than that. When you mix reds and greens, you get yellow and oranges. Add some blue, and you get whites. So our colors will always have some share of red, varying from a bright red to paler or more yellow shades. Of course, if you want green flames, trying swapping the "gamma" variable into green and "gamma" times 8 into red. You can adjust these values to give you any color of flame you prefer.

j. The last line actually changes "lvl" so that it's ready for the next cycle. This loop will cycle very quickly to create the flame effect. If you want to slow it down, just add a green "pause" block from the "Loops" panel at the end.

k. Give your code a name, such as "Campfire." Click the "Save" icon (the small floppy disk), and then click "Download." A .uf2 file will be downloaded to your computer (Figure 5.5).

l. Use the USB cable to plug the CPX into your computer. Tap the small "Reset" button in the center of the CPX.

m. Look for the CPLAYBOOT drive in File Explorer or Finder. Copy the .uf2 file from your computer onto the CPLAYBOOT drive by dragging it over.

NOTE: The complete code for this project is available at https://makecode.com/_YbiFe449jHcb.

Process: Building the Campfire

For this, we'll use simple materials to create a small campfire that our CPX can light. The most challenging part to find is the motor. You can find motors at hobby stores, order them online from electronics distributers, or even hack an electric toothbrush or toy. Just make sure that your motor is powered by 3 volts; otherwise, this project won't work.

You will also need a small propeller that fits the shaft of the motor. You can usually find propellers in the same place you find the motors. Just be sure that it is not wider than the diameter of the cup you are using. You can make your own from a bit of scrap plastic, such as the top from a take-out container. Playing with blade design is a great engineering activity. Just make sure that the propeller attaches firmly to the motor shaft.

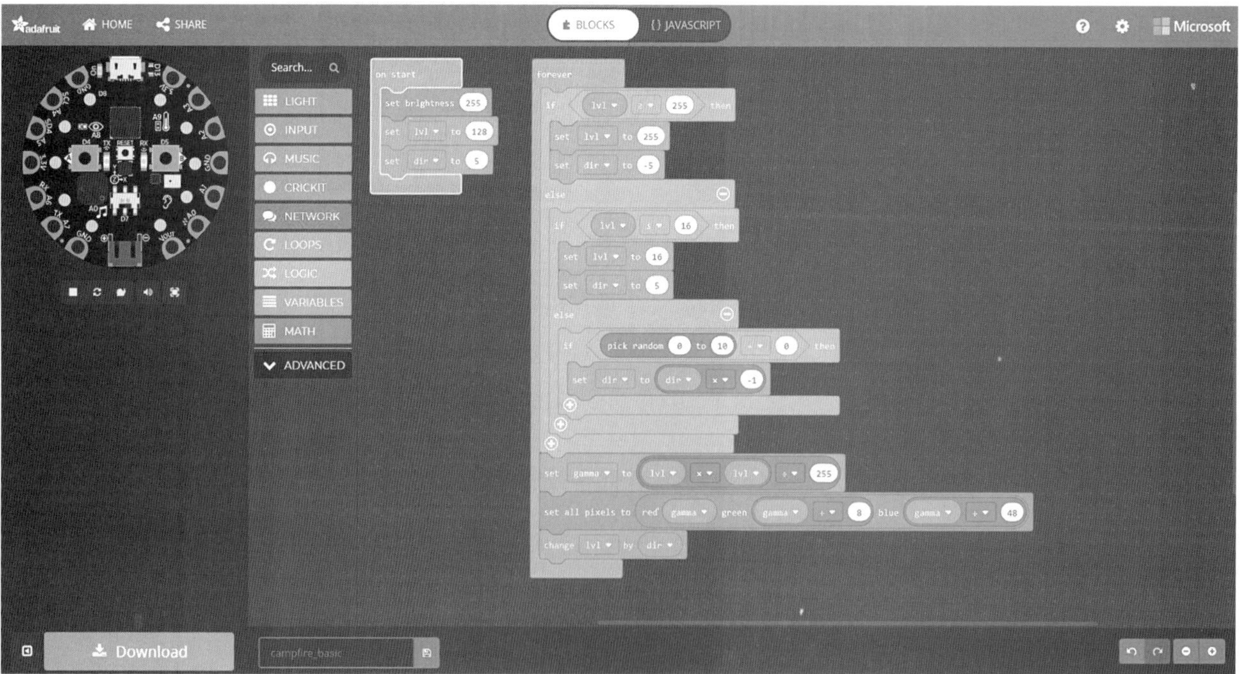

FIGURE 5.5

When selecting a battery pack, make sure that it has wires that are long enough to give you room to work and a switch. The switch is necessary because otherwise your motor will run nonstop. The cable should be about 6 inches long.

1. If you plan to decorate your cup, do so now. You can hot glue stones to the base and twigs around the cup to create a campfire look. Alternatively, you may want to punch holes in the cup to enhance the light show.
2. Put on your safety gear.
3. Use the tape measure to mark 2 inches from the bottom of one of the cups. Use your scissors to trim the cup, reserving the bottom. Then cut a circle in the bottom of the cup, trimming away most of the base but leaving a small ledge, about ⅛ inch (Figure 5.6).

FIGURE 5.6

4. Use the screwdriver or awl to poke a hole in the bottom of the second cup, near the edge. This will be used to thread the battery pack cable into the CPX.
5. Trace the bottom of your motor onto the bottom of the medicine cup. Use the screwdriver or awl to poke a hole in the center of the base. Use the scissors or a craft knife to trim a hole along the tracing you made.
6. Connect one alligator clip lead to the "GND" (ground) contact on the CPX, clipping vertically from below. Connect this to the black lead of your motor.
7. Connect the other alligator clip lead to the "3.3V" contact on the CPX, clipping vertically from below. Connect this to the red lead of your motor.
8. Push the motor through the mouth of the medicine cup and up into the hole just far enough that the base of the motor is higher than the rim of the cup. The medicine cup will form a stand to hold the motor above the CPX. The motor should be secure and not able to shake or fall out of the medicine cup. If needed, secure the cup to the motor with a little hot glue or duct tape. Cut a small slit in the edge of the rim and gently slide the motor leads into it so that they are secure. Attach the propeller to the motor (Figure 5.7).

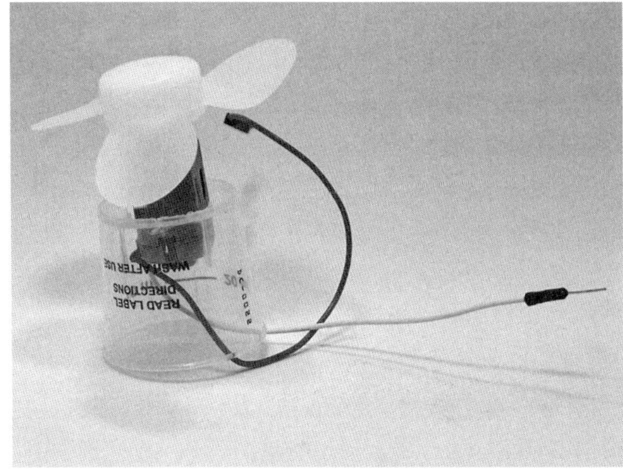

FIGURE 5.7

9. Attach an alligator clip to the CPX on the "VOUT" connector. This is for voltage out. It will power the motor. Clip the other end of the alligator clip to the red lead of the motor.
10. Attach an alligator clip to the CPX on one of the "GND" connectors. This is for ground. This will complete the circuit between the CPX and the motor. Clip the other end of the alligator clip to the other motor lead (Figure 5.8).
11. Turn the second (cut) cup over so that the mouth of the cup is facing downward. Gently push the wires of the alligator clip leads and the leads from the motor below the shelf you have created. If needed, use scissors to remove additional cup material. Lay the CPX with the neopixels facing upward toward the mouth of the uncut cup. Use adhesive foam squares to attach the CPX to the cup. Use squares to attach the medicine cup to the center of the CPX, taking care not to block the neopixels.
12. Feed the battery cable into the hole in the bottom of the cut cup. Make sure that the switch is set to "off," and attach the battery pack to the CPX. Test that everything works (Figure 5.9).
13. Now comes the tough part. Disconnect the battery pack. Feed the connector into the hole made on the side of the uncut cup.
14. Push the cut cup with the CPX and motor into the uncut cup. Feed the cable up under the cut cup and attach to the CPX. Gently push the CPX and motor down into the uncut cup (Figures 5.10 and 5.11).

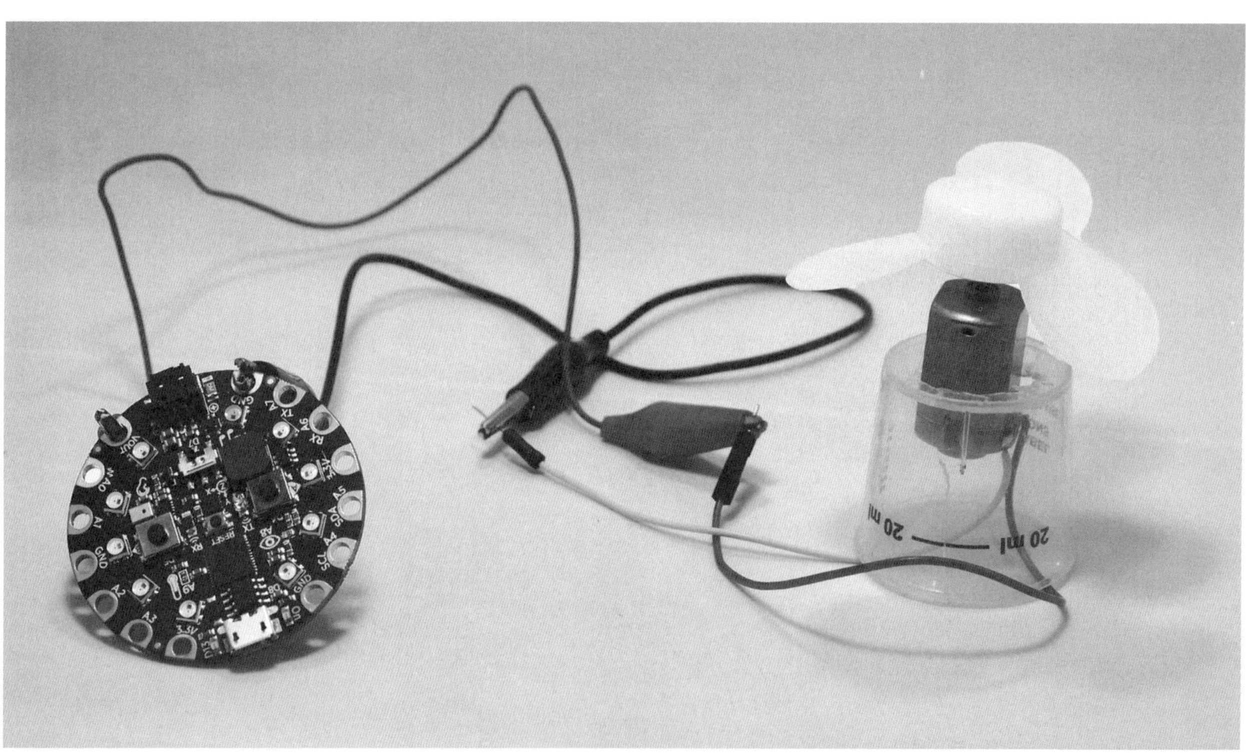

FIGURE 5.8

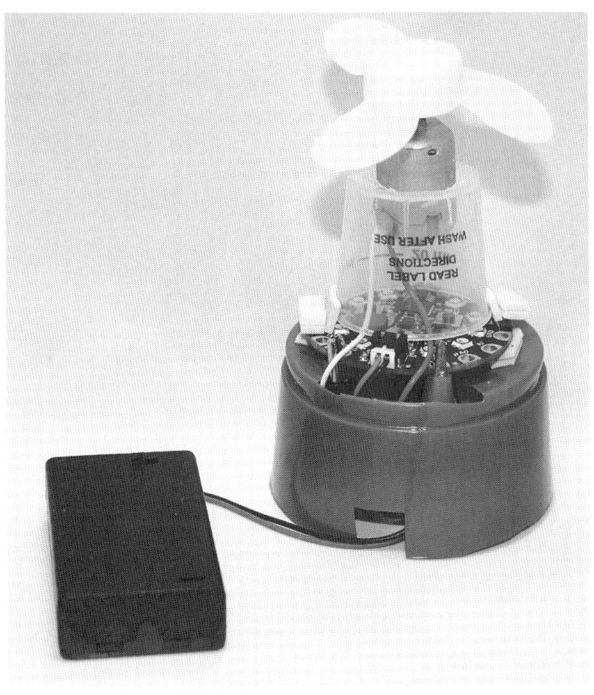

FIGURE 5.9

FIGURE 5.10

FIGURE 5.11

15. Cut paper flames from the tissue paper. Keep the flames under 3 inches in height, varying them by width and height. Thin flames work best. You can trace flame shapes onto the paper or freehand them. It saves time to layer several sheets of tissue paper and cut them all at once. Cut enough flames to go around the circumference of the rim of the cup several times.
16. Using invisible tape or hot glue, attach your flames to the cup and straws. I found it helpful to lay out straight lines of flames, use invisible tape to attach them into a chain, and then add them to the cup of straws.

17. If needed, gently gather your flames toward the center of the cup.
18. Turn the switch to "on." The CPX should create a flickering flame effect while the fan gently moves the tissue paper. If needed, adjust the height of the motor or use scissors to thin the flames to increase movement until the desired effect is achieved (Figure 5.12).

FIGURE 5.12

Take It Further

Try changing the color of the flames by adjusting the "set all pixels to" block in the code. You can also use the capacitive touch sensors on the CPX to change the color by touch. For sample code illustrating these possibilities, visit https://makecode.com/_fXcg4v4srHMa.

PROJECT 4
LED Fireflies and Other Origami Fun

COST: $

MAKE TIME: 15 minutes each

If your campers are anything like mine, at least one will love origami. The ancient Japanese art of paper folding is a fantastic maker activity on its own, offering endless opportunities for adaptation and exploration while also teaching mechanics, mathematics, and more. Add some simple circuits to make your creations light up! Well, that's really taking it up a notch.

This LED firefly is based on a Make Media Maker Camp project from several years ago (available at https://makercamp.com/projects/origami-firefly). I changed the original project to make it easier for young children and those who have not done origami before. This project also uses more commonly found, inexpensive materials than the original. I've also included a light-up origami luna moth and a ninja star, which looks neat thrown in the dark and captured with long-exposure photography (i.e., painting with light) (Figure 5.13).

When doing origami with campers, it's helpful to remind them of a few tips and tricks to make their folding more successful. The first is that they want to make sure to crease their folds well, using a bone folder, a ruler, or a fingernail to make the fold flat and sharp. It is also important to take your time and line up folds as neatly and precisely as possible. Rushing will only yield messy origami that may not work or look as expected. Origami is an excellent activity for campers who need to work on slowing down and focusing on the task at hand. Lastly, remind campers that, after all, it's only paper. If your first attempt doesn't work out, it's easy enough to try again!

FIGURE 5.13

For details on the electronics behind this project, check out the "Light-up Letter Home" in Chapter 4.

Materials

- Origami paper
- 5-millimeter LEDs in various colors
- 3-volt coin batteries (CR2032 or CR2025)
- ¼-inch copper tape with conductive adhesive
- Googly eyes
- Glue sticks
- Invisible tape
- Scissors
- Needle-nose pliers (optional)

Firefly

1. Place the paper in front of you with the points facing in compass directions to make a diamond (Figure 5.14). Create a "mountain fold," folding the top point to the bottom point (Figure 5.15).

FIGURE 5.14

FIGURE 5.15

2. Fold down the left and right flaps so that the top is halfway along the top fold and the corners of the paper point downward and a little away from each other at the bottom. These will be the wings of the bug (Figure 5.16).

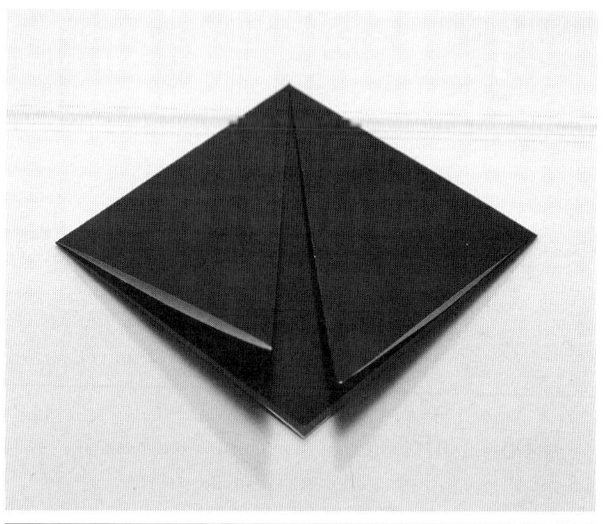

FIGURE 5.16

3. Flip the bug over.
4. Fold the top point down about 1 inch and crease well. Fold the point back up, about three-quarters of the length, to create a pleat. This is the head of the bug (Figure 5.17).

FIGURE 5.17

5. Fold in the left and right sides, making the fold at an angle and overlapping the head pleat by a little bit so that the body of the bug is more narrowed and tapered. The flaps from the front should meet in the center of the back and even overlap by a bit (Figure 5.18).

FIGURE 5.18

6. Flip the bug over, and crease down the center (Figure 5.19).

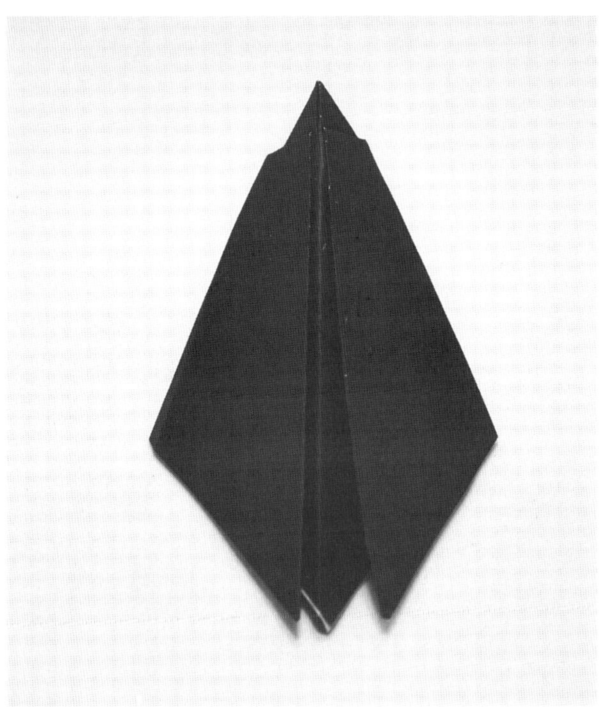

FIGURE 5.19

7. Using a glue stick, attach googly eyes to the head, if desired.
8. Select a yellow or white LED. A LED has two legs, or leads. The long one is the positive (+) lead. The short one is the negative (–) lead. It's important to know which is which. It can help to take a permanent marker and color the negative lead black. Test your LED before going any further. Slip the LED over the battery so that the positive lead lies against the positive side of the battery (marked with a +) and the negative rests against the other side. The LED should light up.
9. Before attaching the LED to the card, you will need to bend the leads. Using your fingers or a pair of needle-nose pliers, *gently* spread and bend the leads of your LED. This will make it easier to attach the LED to your firefly. Take care not to break the leads. If you like, you can curl the leads into circles. This will give you better contact with the circuit when you attach it.
10. Lay the LED with the top facing down. The negative lead should be to the right and the positive lead to the left. Use a little invisible tape to attach the "bulb" of the LED to the firefly (Figure 5.20).

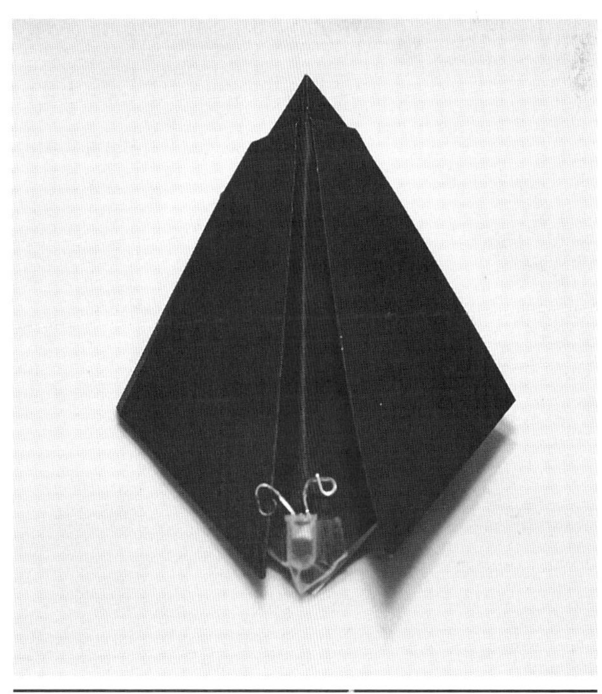

FIGURE 5.20

11. Cut a piece of copper tape that is long enough to go from the negative lead of the LED, wrap around the bottom and reach the spot where the two flaps meet, roughly in the center of the back of the bug. Removing the backing from the tape a little at a time, place the copper tape in place. Do your best not to wrinkle the tape too much or break the tape along the path. If a break occurs, be sure to "patch" it with additional copper tape. Place the negative lead on top of the tape and use additional copper tape to secure it, pressing down firmly to ensure good contact. Do not cover the positive lead with this piece of copper tape!
12. Unfold the right wing of the firefly. Cut a piece of copper tape that is long enough to go from the negative lead of the LED, wrap around the bottom, and travel out onto the inside of the unfolded wing of the bug to roughly the place where it touches the center of the bug when folded. It may help to mark this spot before planning your copper tape. The goal is for the two strips of tape to touch when the firefly is folded. Removing the backing from the tape a little at a time, place the copper tape in place. Do your best not to wrinkle the tape too much or break the tape along the path. If a break occurs, be sure to "patch" it with additional copper tape. Place the positive lead on top of the tape, and use additional copper tape to secure it, pressing down firmly to ensure good contact. Do not cover the negative lead with this piece of copper tape (Figures 5.21 and 5.22).

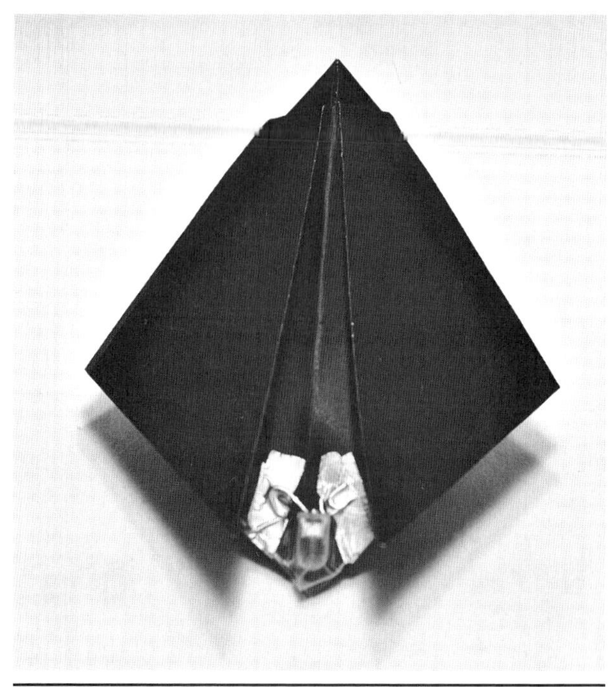

FIGURE 5.21

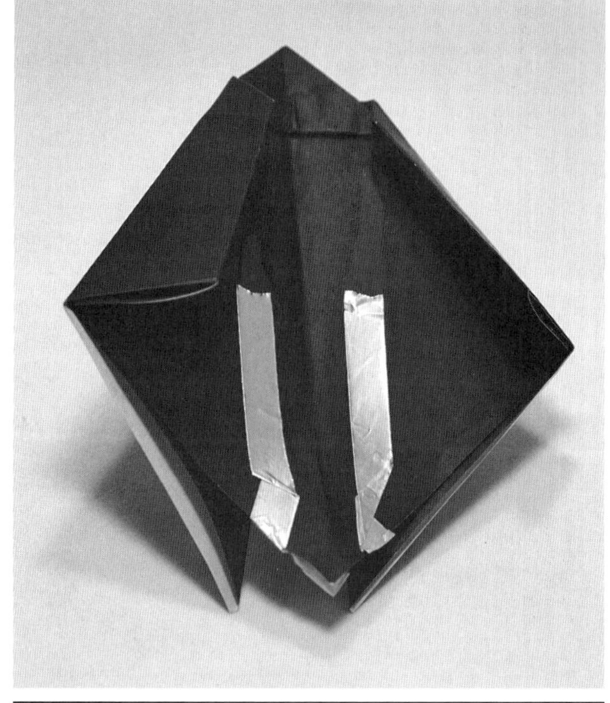

FIGURE 5.22

13. Now it's time to attach the battery. The battery has two sides. The positive side is smooth and marked with a plus sign (+). The negative side is rough and unmarked. Place the coin cell battery with the positive (+) side face *down* on your circuit. Use clear tape to attach it on the edges, but leave metal exposed. If desired, you can fold over a piece of copper tape into a loop, with the sticky side out, and use this to further secure the battery. The goal is to make sure that the battery has good contact with the circuit (Figure 5.23).

14. To light up your firefly, fold the wing back into place, making sure the copper tape touches the negative side of the battery. Hold it there. The LED should light up. Once you've confirmed that your circuit works, use invisible tape to tape the wings together on the back of the firefly. Flip the bug over. Squeeze the bug at the center, and it should light up. If desired, you can use copper tape to attach the positive lead to the top of the battery so that it always lights up (Figure 5.24).

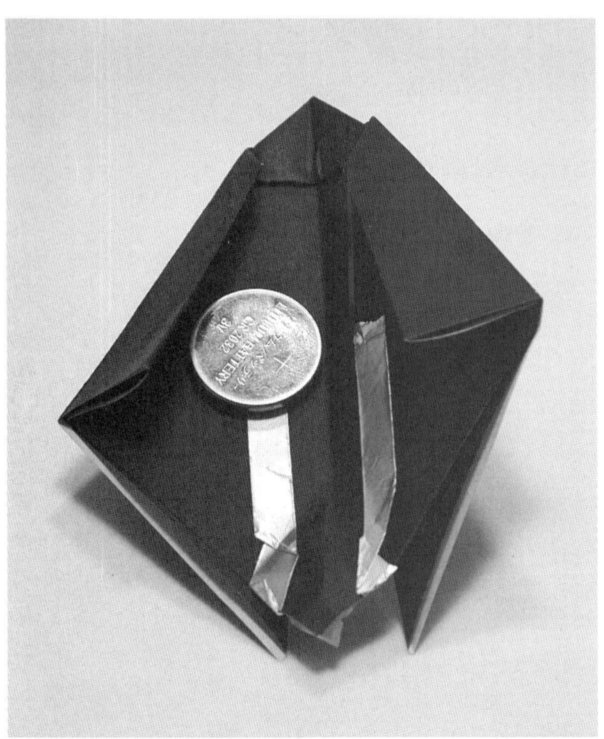

FIGURE 5.23

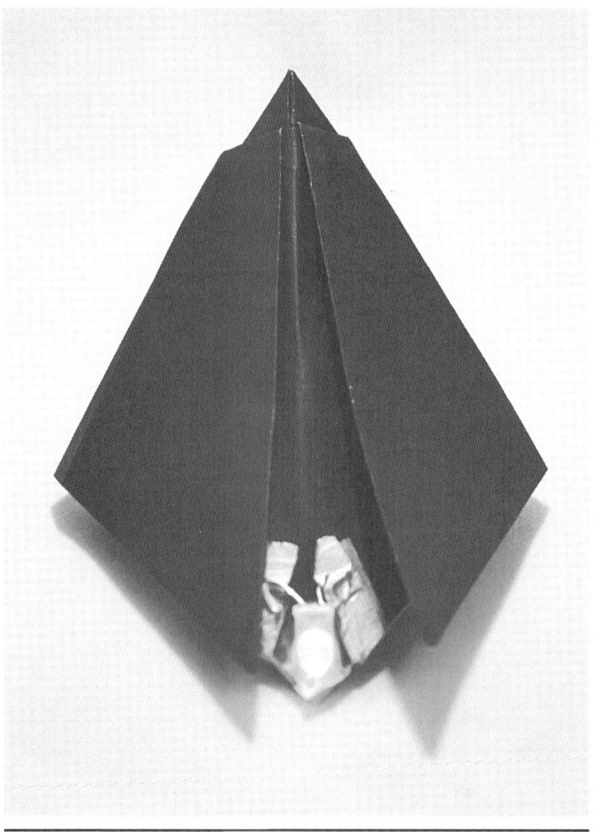

FIGURE 5.24

Luna Moth

Glow-in-the-dark origami paper is available online. How cool is that? This origami is a bit more challenging than the firefly and uses two LEDs wired in parallel rather than series so that you have glowing antennas. Pale-green paper would be the traditional color, but it's your moth, so the sky is the limit.

1. Place the paper in front of you with the points facing in compass directions to make a diamond. Create a mountain fold, folding the top point to the bottom point. Unfold.
2. Turn 90 degrees, and repeat step 1 (Figure 5.25).

FIGURE 5.25

3. With the paper returned to the diamond orientation, fold the top point down to the center of the paper (Figure 5.26). Repeat for the other points. Unfold (Figure 5.27). Flip the paper over and orient as a square (Figure 5.28).

FIGURE 5.26

FIGURE 5.27

FIGURE 5.28

4. Fold the top to the middle line (Figure 5.29). Repeat for the bottom (Figure 5.30). Turn 90 degrees and repeat (Figures 5.31 and 5.32).

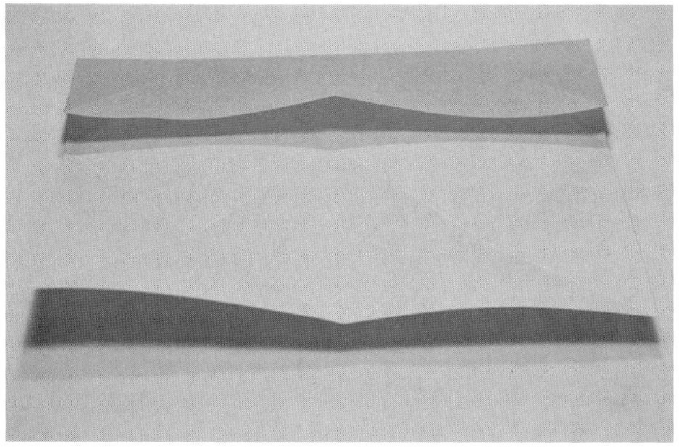

FIGURE 5.29

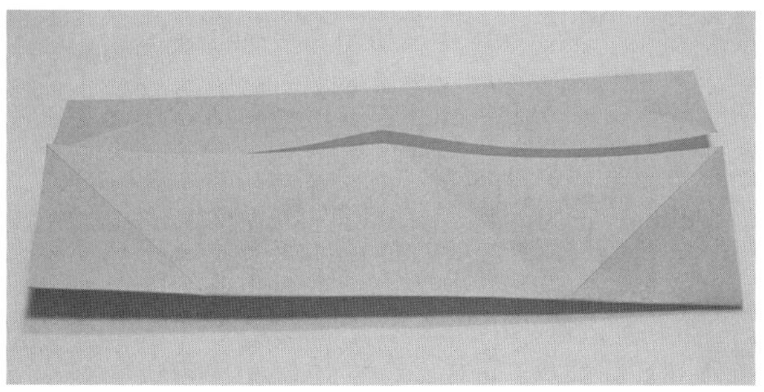

FIGURE 5.30

FIGURE 5.31

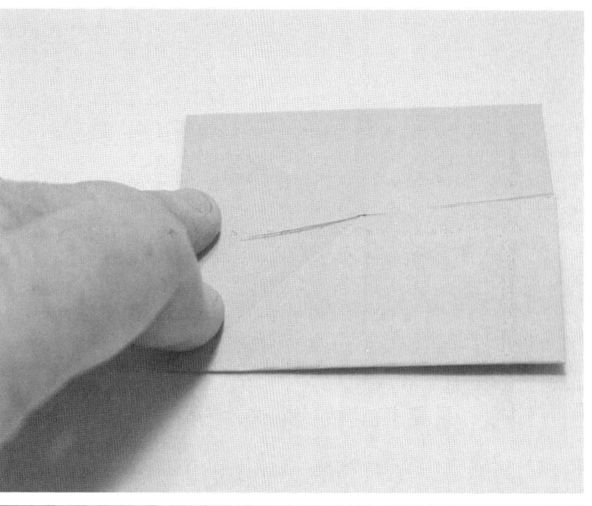

FIGURE 5.32

5. Open the top folds. Take a point toward the center and fold it downward, reversing the folds and aiming the point outward and to the side. Do the same for the other side (Figure 5.33). Then do the same for the bottom. At this point, you should have a six-sided shape (Figure 5.34).

FIGURE 5.33

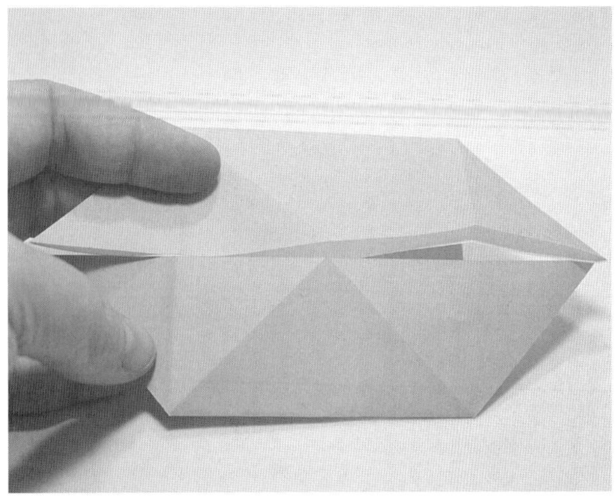

FIGURE 5.34

6. Rotate it 90 degrees. Fold the points downward and out to the side, along the diagonal creases. Flip the paper (Figure 5.35).

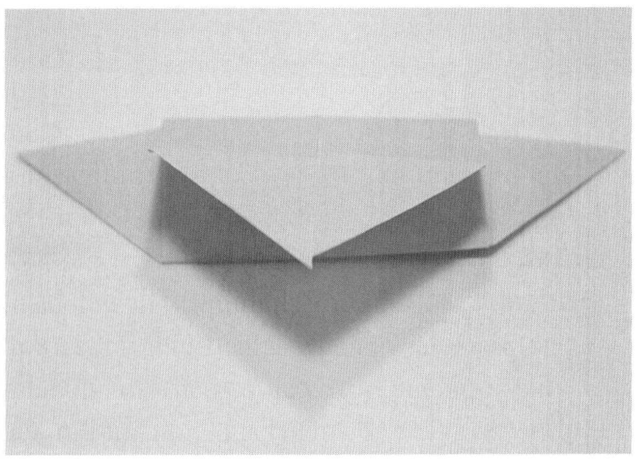

FIGURE 5.35

7. Fold the top down to meet the bottom of the square (Figure 5.36).

FIGURE 5.36

8. Fold the right side over along the middle, creating an isosceles triangle (Figure 5.37).

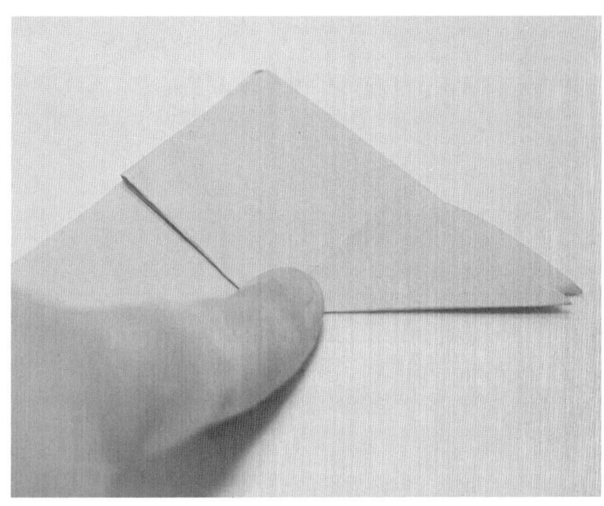

FIGURE 5.37

9. Holding the top of the triangle, at the right angle, fold the other points upward and out, forming wings. The part you are holding will be the body of the moth. It should be about an inch in depth at the center, wider toward the top, and coming to a point slightly past halfway along the moth. Crease well. If desired, trim the edges to curve them like a luna moth's wings (Figure 5.38).

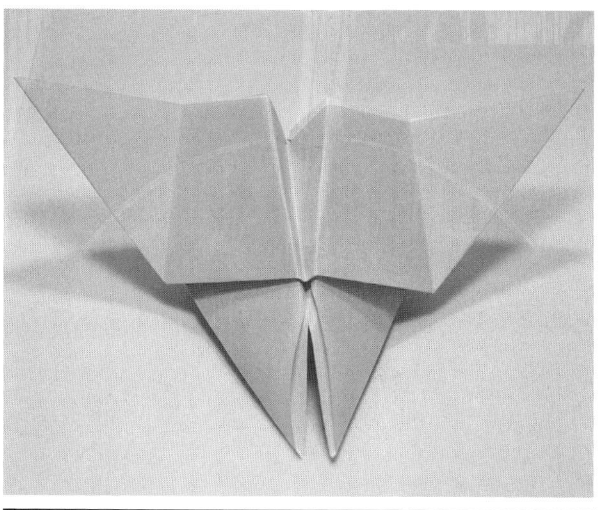

FIGURE 5.38

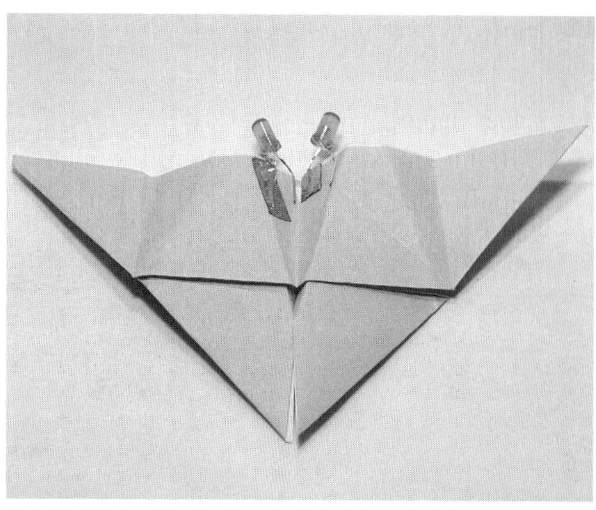

FIGURE 5.39

10. Flatten the moth. Inside the folded body, arrange the first LED on the left so that the positive lead is to the left and the negative lead is to the right, with the LED roughly centered between the center and diagonal fold. Use copper tape to attach each lead, taking care not to let the tape for the negative and positive leads touch. This will short the circuit.

11. Place the second LED with the negative lead to the left and the positive lead to the right (the opposite of the first LED), with the LED roughly centered between the center and diagonal fold. Use copper tape to attach each lead, taking care not to let the tape for the negative and positive leads touch. This will short the circuit. You do want the two positive leads to be able to touch when the moth is folded together. The negative leads should also touch when the moth is folded, but the negative and positive leads should not touch at any point. Place a longer piece of tape on the left-most positive lead of the first LED and the left-most negative lead of the second LED (Figure 5.39).

12. Place the battery. (See step 12 of the firefly process for battery basics.) Place the positive side of the battery face *down* on the left-most positive LED lead, taking care *not* to touch the negative lead (Figure 5.40).

FIGURE 5.40

13. Adjust your tape and folds as needed to accommodate the battery. If desired, you can fold a piece of copper tape into a loop with the sticky side out and use this to further secure the battery. Test the circuit by pressing the body together at the top. The positive and negative leads should have good contact, and the moth's antennas should light up. Adjust the LEDs by bending them to separate the lights (Figure 5.41).

FIGURE 5.41

14. Once you have confirmed that it works, use a small amount of invisible tape to secure the battery as needed. Make sure that the copper tape can still make contact with the metal of the battery.
15. Gently bend the LEDs outward so that they don't touch. Flap your moth's wings to activate the circuit!

More Light-up Bugs

One of the best collections of insect origami projects I have found online is by the Origami Resource Center at https://www.origami-resource-center.com/origami-insects.html. Campers interested in exploring more challenging origami projects will find many ideas for bees, grasshoppers, dragonflies, beetles, and much more there. Experiment with ways to add light to your creations using what you have learned.

You may also want to check your local library for these books:

- *Origami Insects and Their Kin*, by Robert J. Lang (Dover Publications Inc., 1995)
- *Paper Bugs!: Fold Origami Insects and Other Invertebrates: Butterflies, Dragonflies, Bees, Roaches, Snails, and More*, by Carmel D. Morris (CreateSpace Independent Publishing Platform, 2012)
- *Bugs in Origami*, by John Montroll (Dover Publications, 2013)
- *Origami Masters Bugs: How the Bug Wars Changed the Art of Origami*, by Jason Ku and Sebastian Arellano (Race Point Publishing, 2013)
- *Origami Insects: Easy and Fun Paper-Folding Projects*, by Anna George and Diane Craig (Super Sandcastle, 2016)

Another fantastic light-up animal origami creation was offered by Kathy Ceceri in *Make Magazine* in September 2017, based on a project in her book *Paper Inventions: Machines that Move, Drawings that Light Up, and Wearables and Structures You Can Cut, Fold, and Roll* (Maker Media, Inc., 2015). Her origami jumping frog with LED eyes is a great project for older campers. The instructions are available at https://makezine.com/projects/fold-up-jumping-origami-frog-led-eyes/.

Ninja Star

This simple throwing star was originally based on a project from *Mini Weapons of Mass Destruction: Build and Master Ninja Weapons*, by John Austin (Chicago Review Press, 2014). Austin's books should be in every makerspace because they offer a lot of inexpensive, engaging projects. I've been making my own version of the throwing start for years, and when I explored painting with light a few years ago, it was no surprise that my campers' creativity led to LEDs being added to everything, often with amazing results.

This project is best with two-color reversible origami paper. It is the most challenging of the four projects. You will be working with two pieces and mirroring the folds on them. You will need four LEDs for this project. If you have access to 3-millimeter LEDs, use them rather than the more common 5-millimeter LEDs to reduce the weight on the star. Even better, Chibitronics' stick-on LED stickers will make this project even easier and lighter. This project doesn't have a switch, so it will stay lit all the time. As a result, the battery won't last as long as those in the other projects.

If you would like to experiment with painting with light, you will need a long-exposure camera app for your phone or tablet. Both Android and Apple have free versions available, such as Low Shutter Cam and LongExpo. You can also use a DSLR camera set for long exposure. For additional instructions, see https://makercamp.com/project-paths/paint-with-light/.

1. Select two squares of origami paper. Fold each square of paper in half, and crease well. Unfold (Figure 5.42).

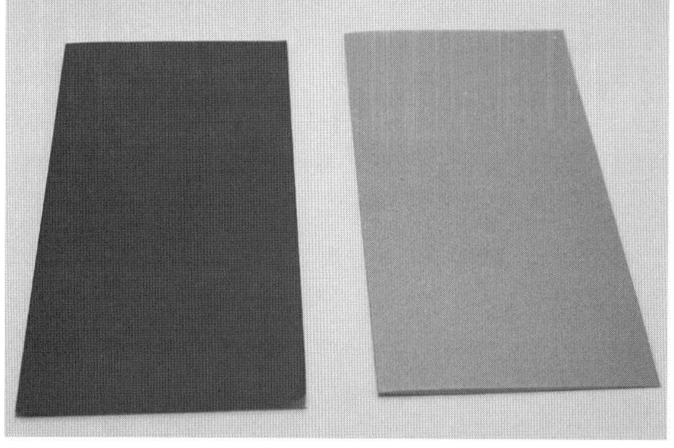

FIGURE 5.42

2. Fold the outer sides of each paper into the center line, and crease (Figure 5.43). Fold along the center line again to form a rectangular strip (Figure 5.44).

FIGURE 5.43

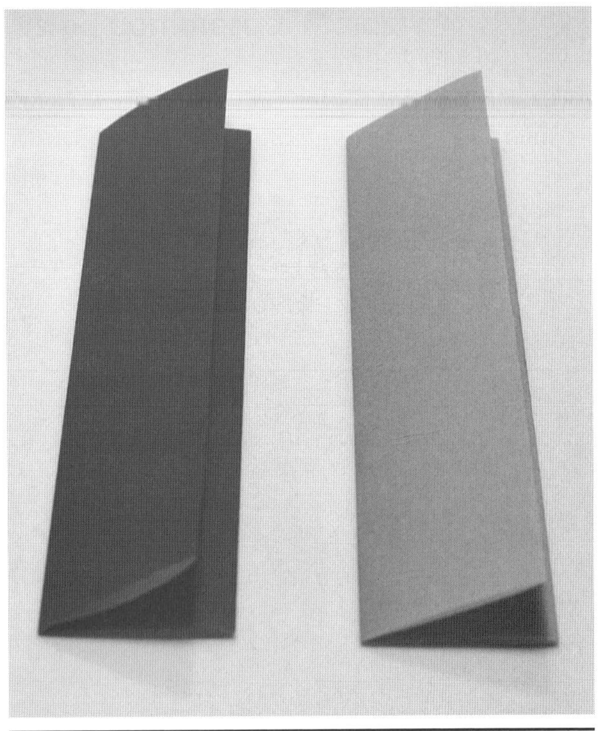

FIGURE 5.44

3. Fold both pieces in half top to bottom, crease well, and unfold (Figure 5.45).

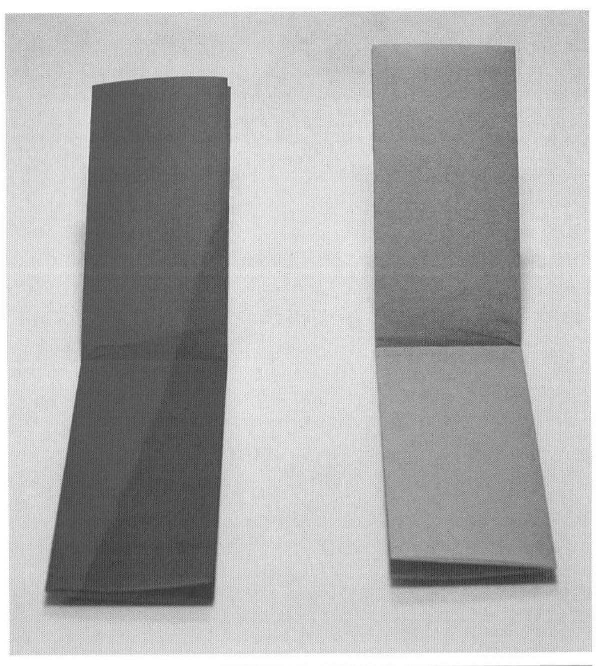

FIGURE 5.45

4. Take one piece and fold downward from the top and diagonally to the right along the midway crease you created in the last step. Repeat for the other rectangular piece, but fold to the left instead. Crease well (Figure 5.46).

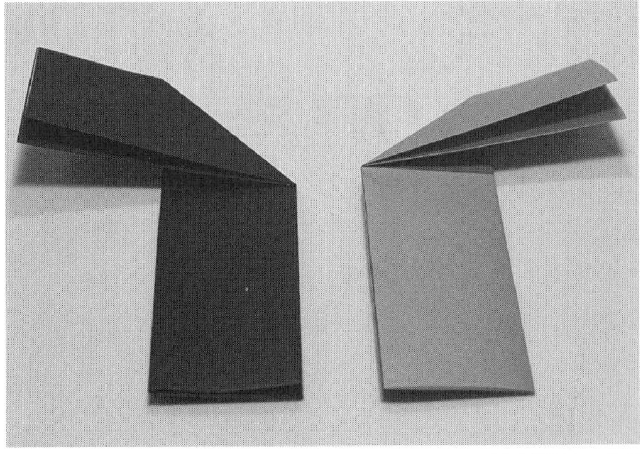

FIGURE 5.46

5. For each piece, fold the bottom up and diagonally in the opposite direction of the top fold, again lining it up along the midway fold (Figure 5.47).

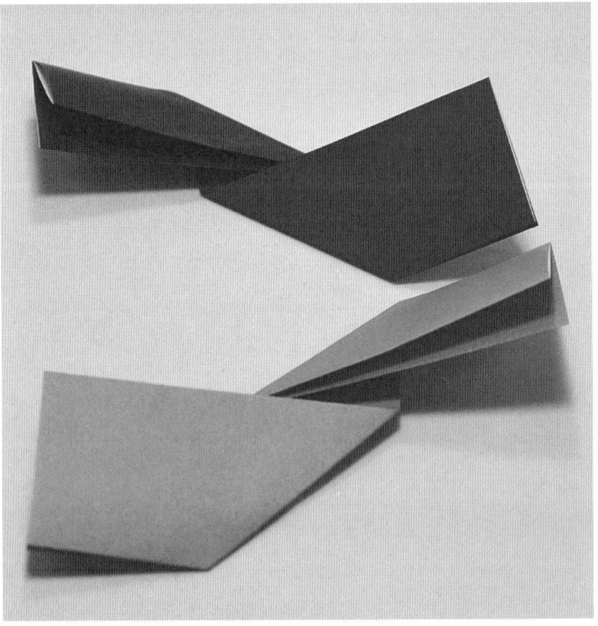

FIGURE 5.47

6. Flip each piece horizontally so that the top folds point toward one another and the back of the piece now facing you.
7. On each piece, fold the upper outside edge into the opposite corner. Repeat for the bottom outside corner.
8. On each piece, take the outermost top point (the one you created in the last step), and fold it in straight across to the outside edge of the piece. Do the same with the bottom point. You should now have two mirrored parallelograms (technically two rhombuses). Crease well, and unfold (Figure 5.48).

FIGURE 5.48

9. Turn your right piece over horizontally. Place it on top of the left piece. No points should overlap (Figure 5.49).

FIGURE 5.49

10. Starting with the bottom-left point, fold in straight over the other piece, and tuck it into the flap opposite the point. Repeat with the upper-right point (Figure 5.50).

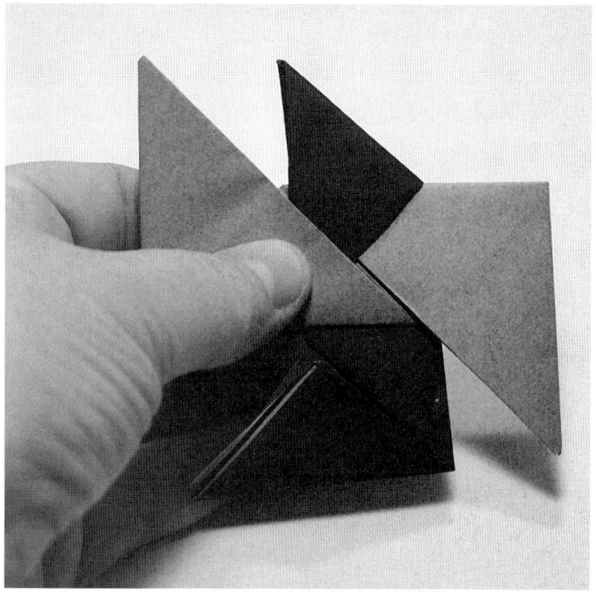

FIGURE 5.50

11. Flip the star. Repeat step 10.
12. Flatten your ninja star, making sure that the points are tucked and the creases are sharp (Figure 5.51).

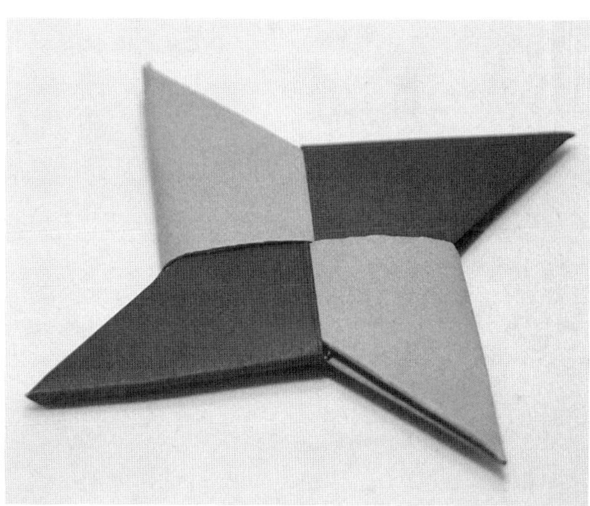

FIGURE 5.51

13. Prepare four LEDs as described in steps 7 and 8 of the firefly process.
14. Orient your throwing star so that the points align with the compass. Place your first LED with its top facing the point at north. Place the positive lead to the left and the negative lead to the right. Use a small piece of invisible tape to secure the "bulb" of the LED to the star.
15. Rotate the star 90 degrees. Place a LED with its negative lead to the left and its positive lead to the right. Attach as above. Rotate 90 degrees. Place a LED with its positive lead to the left and the negative to the right. Attach as above. Rotate 90 degrees. Place a LED with its negative lead to the left and its positive to the right. Rotate 90 degrees (Figure 5.52).

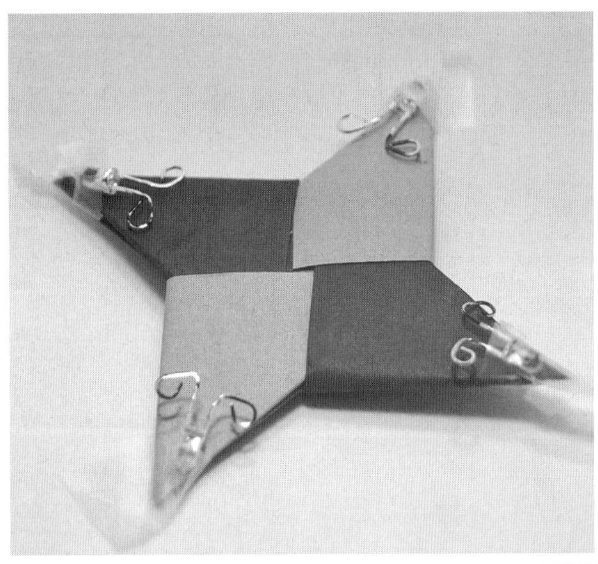

FIGURE 5.52

16. Refer to Figure 5.53 to lay down the copper tape. Use a piece of copper tape to connect between the north and east negative leads. Use a piece of copper tape to connect between the south and west negative leads.
17. Use a piece of copper tape to connect between the west and north positive leads. Use a piece of copper tape to connect between the east and south positive leads.
18. Measure and cut a piece of tape that can wrap around the throwing star. Use the tape to connect the strip of tape connecting the west–north positive leads to the east–south positive leads by wrapping it around the back of the star. Place a piece of copper tape from the east–south positive lead into the center of the star (Figure 5.53).

FIGURE 5.53

19. Review the battery basics covered in step 12 of the firefly project. Place the battery positive face (+) *down* onto the center of the star.
20. Place a piece of copper tape between the east and north negative leads (or the copper tape attached to the negative leads), taking care not to touch the battery or any of the positive leads or copper tape. Use a piece of copper tape to connect the north negative lead to the top of the battery.
21. Press down for good contact to test the circuit. All the LEDs should light up. If not, troubleshoot, looking for breaks in the copper tape, bad connections, or short

circuits. If everything lights up, use invisible tape to securely attach the battery and LEDs as needed. You can remove the tape on the bulbs of the LEDs at this point. You may want to secure the battery with invisible tape (Figure 5.54).

22. Test throwing your star. Hold the star horizontal to and parallel with the ground. Flick your wrist and throw the star. It should rotate and fly horizontally. You can also try throwing it at a perpendicular angle to the ground.
23. Try turning off the lights and using a long-exposure app on your smartphone to take pictures of the lit throwing star in motion.

FIGURE 5.54

PROJECT 5
Spooky Ghost Photos

COST: $$$

MAKE TIME: 1 minute each

What camping trip would be complete without at least one or two spooky ghost stories? Even if your campfire is powered by LEDs, it can be fun to gather around and share classic spine-tingling stories. Believe it or not, confronting scary ideas while in a safe place and with people they trust can be an empowering experience for kids that can help them confront real-life scary moments with more confidence.

To add a maker twist to the ghost stories, why not make storytelling a collaborative game? The simplest way to do this is to have one camper start the story and then have every other camper in the circle add to it, until you reach an ending point. Or you could pair campers up to complete Halloween Mad Libs and then share the results.

Of course, your campers may want photographic evidence of their ghost adventures. If you have a DSLR camera, it's actually very simple to create your own ghost photos. Not only is it a ton of fun to make these spooky pics, but it also gives you a great opportunity to introduce critical thinking. Once campers see how easy it is to create fake photos, they may think twice about what deceptive media they see in their everyday lives. You can have some great discussions about how we can evaluate the validity of information we see every day.

This project also gives campers an opportunity to explore some of the basic settings on the camera, thereby introducing basic photography techniques. You may want to share a short video with campers about DSLR basics before you begin. YouTube has many great options.

Materials

- A DSLR camera
- A tripod for the camera
- A room with dim lighting
- Old-fashioned costumes or spooky props (optional)
- A spray water bottle (optional)
- A glow stick (optional)

Process

1. Mount your camera on a tripod.
2. Adjust the settings for a long exposure of at least 10 seconds. (Refer to your camera's user's manual if you are unsure how to do this.) This will leave the lenses of the camera open so that you can essentially layer many pictures on top of one another and will give you the time to create your ghost.
3. Make the camera aperture small. Start with f/22. A small aperture limits the amount of light that can enter the lens and limits the depth of field, creating a more fuzzy image, which we want for this project. The higher the f-stop number, the smaller the aperture. Experiment until you get results you like.
4. Adjust the ISO level to high. You want to give the camera time to take in more light. Try something in the 400 to 800 range.
5. Have a camper sit or stand very still in front of the camera. Click the camera to release the shutter. Wait 5 seconds, and have your camper either duck out of the shot or move quickly out of the scene to the left or right. Wait until the camera completes the shot. By removing your camper before the shutter closes, you are creating the translucent, ghostly image (Figure 5.55).
6. Check your photo, and adjust the setting as desired.

FIGURE 5.55

7. If you want to add ghostly orbs, simply spray water from the bottle into the shot. Want a scary streak? Try tossing a glow stick through the scene.

NOTE: If you don't have a DSLR camera, you may still be able to create ghost photos. Try the "Fireworks" or "Night" settings on your camera or smartphone photo app. You can also try using a long-exposure photography app.

Take It Further

Some campers may want to import their photos into Photoshop or GIMP to continue editing. They can use the transparency features to layer photos, creating ghostly effects. This is a fun introduction to graphic arts, if you have access to a computer.

PROJECT 6
Making Musical Instruments

Music is a universal language that combines art, science, and mathematics into so many creative expressions that span across the globe and across all ages. For this reason, I feel that Maker Camp really should include music in some way. From making musical instruments to making up a camp song, most campers enjoy the energy of making music together.

You don't need a lot of money to make fun musical instruments with campers. In addition to letting campers express themselves by decorating and using their instruments, there is a lot of amazing science to sound. Make the most of the opportunity. When banging on your drum, talk about why there is no sound in space. When making rubber band guitars, experiment with pitch. The possibilities for crafting and engineering are endless.

If you have access to technology such as a Makey Makey or Circuit Playground, it can be a ton of fun to introduce technical concepts. Using the classic "Banana Piano" as a jumping-off point, challenge campers to create other fun instruments and code them. Campers who can read music may enjoy translating it into code that can be played back in a wearable circuit. Block languages such as Scratch and MS MakeCode make it easy for every camper to play with his or her own sound lab.

Coffee Can Drum

COST: Free–$

MAKE TIME: 15–20 minutes

The heart of sound is vibration, so it only makes sense that drums are some of humans' first instruments. Encourage campers to test and try a variety of materials for the drum body (the shell) and the head (or batter) to create different sounds. Experiment with different heights for the drum and different tensions on the head.

Materials

- A metal or plastic coffee can or cardboard oatmeal container
- Various head materials, such as wax paper, paper bags, copy paper, construction paper, assorted fabrics, plastic wrap, etc.
- Rubber bands
- Drumstick items such as pencils, chopsticks, knitting needles, etc. (optional)
- Markers, stickers, feathers, and other materials for decorating

Process

1. Select a container as your drum shell. Decorate it as desired.
2. Select a material for your drum head. Stretch it tightly across the top of your shell. Use several rubber bands to secure the head.
3. Test your drum. Try different materials until you get the sound you want (Figure 5.56).

FIGURE 5.56

Water Bottle Maracas

COST: Free–$

MAKE TIME: 15–20 minutes

This is a great way to experiment with different sounds while upcycling common materials. For added fun, have campers create their maracas from common materials and then cover the water bottles with construction paper. Challenge campers to figure out what is inside the water bottle. Younger campers love this game.

Materials

- 8-ounce disposable water bottles, cleaned and dried, with top (or tape to cover)
- Toilet paper tubes
- Masking tape or colorful duct tape
- Construction paper or other materials for decoration
- Several funnels
- Materials to fill the maracas such as dried beans, dried lentils, dried corn, rice, seeds, assorted beads of various sizes, sequins, pencil top erasers, marbles, paper clips, brass fasteners, nuts and bolts, pebbles, sand, small pasta, etc.

Process

1. Using the funnel, fill the water bottle about halfway with your material of choice. Cap the bottle, and test the sound. When you are happy with the sound created, tape the cap on to secure it.
2. Place the toilet paper tube over the neck of the water bottle and secure it well with tape.
3. Decorate as desired (Figure 5.57).

Alternatives

If you don't have water bottles on hand, tape together two paper cups instead. Or staple together two small paper plates or bowls.

FIGURE 5.57

Paper Plate Tambourine

COST: Free–$

MAKE TIME: 15–20 minutes

No camp band would be complete without the jingly jangly joy of a tambourine. Jingle bells are the obvious choice here, but you can get creative with any metal objects you have on hand, such as bottle caps or jar lids. If needed, use a nail and hammer to punch holes in advance, tapping down and taping over any sharp corners. Note that you will need two caps for each spot on the tambourine so that they can clatter against each other.

Materials

- Sturdy lunch- or dinner-sized paper plates
- A handheld hole punch
- Ribbon, yarn, or string
- Jingle bells in various sizes
- Scissors
- Pencils
- Markers, stickers, feathers, and other materials for decorating

Process

1. Using a pencil, make marks evenly spaced every 2 inches apart around the edge of a plate. (You can use your thumb to measure or get fancy with an actual ruler.)
2. Using the hole punch, make a hole at each mark.
3. Cut pieces of ribbon, yarn, or string 4 inches long, one for each hole on your plate.
4. Using the ribbon, tie a jingle bell to each hole, making sure that the knot is secure.
5. Decorate as desired.

Rain Sticks

Cost: Free–$

Make time: 15–20 minutes

Is there anything as magical as the sound of a rain stick? There is a reason: this is a classic. Here again you have many opportunities for experimentation. Bring in not just paper towel rolls but also wrapping paper rolls, mailing tubes, and toilet paper rolls. Tape multiple columns together for longer rain sticks. Try a variety of materials in the rain sticks to see how the sound changes. Vary the number of and placement of barriers inside the stick to create different effects.

Materials

- Tubes from paper towels, toilet paper, or wrapping paper
- Toothpicks, bamboo skewers, wooden dowels, or finishing nails roughly the width of your tube to act as barriers to the objects moving within the tube
- Awls, screwdrivers, pens, or drills as needed to poke holes through your tubes
- Leather work gloves (optional, but suggested)
- Colorful duct tape
- Scrap cardboard
- Scissors
- Hot glue (optional, but suggested)
- Materials for inside the rain sticks, such as dried beans, dried lentils, dried corn, rice, seeds, assorted beads of various sizes, sequins, pencil top erasers, marbles, paper clips, brass fasteners, nuts and bolts, pebbles, small pasta, coins, etc.
- Markers, stickers, feathers, and other materials for decorating

Process

1. Use your awl (or other tool) to poke holes in the tube.
2. Push toothpicks or other materials through the holes. The more the better. Secure each with tape or hot glue on the outside of the tube.
3. Place the end of the tube against the scrap cardboard. Trace the outside of the tube twice. Cut circles from the cardboard. Place one circle against an end of your tube. Secure with tape or hot glue.
4. Turn the tube over so that the open end is facing up. Add your desired materials to the inside of the tube. Place the other cardboard circle against the open end. Secure it temporarily with your hand or tape.
5. Test the sound of your tube by gently turning the tube over. Add or remove materials from inside the rain stick or add more toothpicks as barriers.
6. Once you are happy with the sound, secure the second cardboard circle with tape or hot glue.
7. Decorate as desired (Figure 5.58).

FIGURE 5.58

Cardboard Box Guitar

COST: Free–$

MAKE TIME: 15–20 minutes each

Another classic, the cardboard box guitar gives campers the opportunity to play with pitch. Make sure that you select rubber bands of different thicknesses and lengths so that your campers can experiment. Some suitable boxes for this project include sturdy cereal boxes, cracker boxes, shoe boxes, tissue boxes, diaper wipe boxes, and cardboard mailers. If you have access to old cigar boxes, they make very nice guitars.

A neck and head of a guitar are not technically necessary for this project, but campers may enjoy fashioning them out of paper towel tubes or scrap cardboard. Similarly, you can add a nut and/or saddle on your guitar. A nut helps keep the space of guitar strings and is usually placed near the head of the guitar. You can notch your box to create a nut or use a piece of notched corrugated cardboard glued in place. A saddle is usually placed at the top of the body and can move to adjust the sound of the guitar by varying the "action" of the strings by adjusting their height from the surface of the body. An old pencil, pen, or marker works well.

Older campers may enjoy building their own guitar bodies. You may want to supply a guitar body template to make the process easier. Then campers can trace two surfaces from corrugated cardboard and use a long length of chipboard to form the sides. Hot glue will make it easy to attach the pieces together. A neck and head can be built from cardboard tubes or layers of corrugated cardboard. Older campers may also enjoy playing with various thicknesses of fishing line as strings, attaching them to pushpins glued into the head.

Materials

- A box for the body of the guitar
- Rubber bands of various sizes and thicknesses
- Scissors, box cutters, craft cutting knives, or serrated cardboard cutters
- Leather work gloves (optional, but suggested)
- Goggles (optional, but suggested when working with rubber bands)
- Colorful duct tape
- Old pencils, pens, or markers for the saddle
- Paper towel tubes for the neck, if desired
- Markers, stickers, feathers, and other materials for decorating

Process

1. Cut a circle about a third of the way from the bottom of the box. (If your box already has a hole, simply use that or modify it as needed.) I suggest cutting the hole smaller and widening it if needed. You can't really make the hole smaller once it's cut.
2. Stretch rubber bands of various thicknesses over the length of the box, positioning them over the hole in the body. Secure them with duct tape once you are happy with the sound.
3. If desired, slide your choice of saddle into place under the rubber bands at the top of the body. Move it up and down to see how this affects the sound.
4. If adding a paper towel tube neck, cut ½-inch slits into the bottom of the tube every half inch or so. Bend the cardboard outward to create flaps. Using tape, attach the neck to the top of the guitar. If your rubber bands are long enough, you may be able stretch them from the base of the body to the top of the paper towel tube. Use scissors to make notches at the top of the neck to seat the rubber bands securely, and tape them in place.
5. Decorate as desired (Figure 5.59).

FIGURE 5.59

Take It Further

If your campers can't get enough of making musical instruments, check your local library for *Musical Inventions: DIY Instruments to Toot, Tap, Crank, Strum, Pluck, and Switch On*, by Kathy Ceceri (Maker Media, Inc., 2017). It includes a variety of mechanical and digital projects. Older campers or those inclined toward music could happily spend hours exploring this book. *The Big Book of Makerspace Projects: Inspiring Makers to Experiment, Create, and Learn*, by Colleen Graves and Aaron Graves (McGraw-Hill Education TAB, 2016), also includes a chapter on making musical instruments, including Popsicle stick kazoos and a playground pipe organ made of PVC.

PROJECT 7
Music with the Makey Makey

COST: $$

MAKE TIME: 10-15 minutes

If you have access to a Makey Makey and a computer, you can take your music making into the modern age and code your creations. The Makey Makey is a low-cost tool that allows you to connect any conductive material to a computer and use it as you would the arrows, enter, and click. One of the most famous introductory Makey Makey activities is the "Banana Piano," which uses common bananas as keys for a simple six-key piano that is coded with Scratch on the computer.

Materials

- A bunch of bananas, plus assorted other fruit and vegetables
- A Makey Makey
- 7 alligator clip leads
- A computer with a USB port and access to the Internet

Process

1. Connect the alligator clips to the four directional arrows, enter, and click locations on the Makey Makey board. Attach an alligator clip lead to the ground location.
2. Lay out the bananas on a table. Attach each of the other ends of the alligator clip leads to a banana, making good contact with the skin.
3. Attached the red USB cord to the Makey Makey, and plug it into the USB port of your computer. The green power light should glow on the Makey Makey.
4. Open your internet browser. Go to the Makey Makey banana piano website at https://apps.makeymakey.com/piano/.
5. Hold the ground clip in your hand, touching your skin, and then touch a banana. A note should play. Play several notes or a short song. Invite campers to try it. Does it work without holding the ground clip?
6. Tell campers that they will all become part of the circuit. Have a camper hold the ground clip in his or her hand, touching skin. Have campers hold hands. Have the last camper in the chain touch a banana. A note should still play. Why?
7. Give campers other fruits and vegetables or other household materials, and have them test to see what materials work best. Can they categorize materials as good conductors, poor conductors, and insulators?
8. Go to the Scratch website, search for "Makey Makey Piano," and select one of the options (or try https://scratch.mit.edu/projects/90316728/).
9. Click on "See inside." To change the sound made by each key, click on the sprite (the picture icon) for each key. Sound blocks are purple. For the code mentioned earlier, campers will read something like "play note 60 for ½ beat." Try changing the numbers to see what happens. Or pull that block out

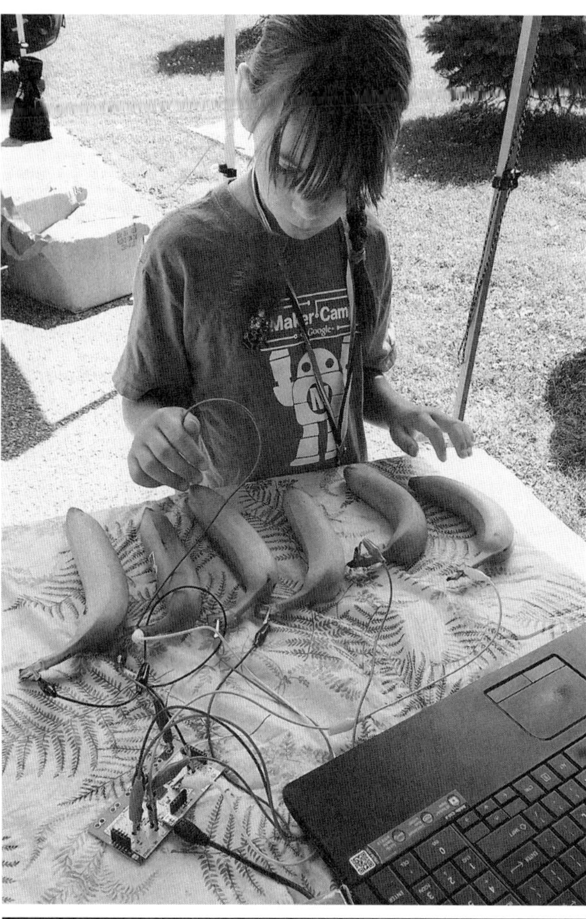

FIGURE 5.60

and add a different block, such as "play sound" or "play drum" (Figure 5.60).

10. If desired, use cardboard and copper tape or aluminum foil to build piano keyboards, drum pads, simple guitars, or other creations to combine with the Makey Makey. See "Light-up Letter Home" in Chapter 4 for information on creating paper circuits. Chapter 4 also has additional Makey Makey projects.

PROJECT 8
Coding a Robot Drum Circle

COST: $$

MAKE TIME: 45–90 minutes

If you have access to a computer lab, this is a fun collaborative coding project. It uses the newly introduced Scratch 3.0 because this easy-to-learn drag-and-drop coding language works on both computers and tablets such as the iPad, making it wonderfully mobile. Once everyone has coded their robot drummer, why not take campers outside to create some digital music? If you are in the lab, consider using the Makey Makey with campers to create simple drum pads that work with their code.

Instructions below are given for a computer using a keyboard and mouse. Some changes may be needed if you are working on a tablet.

Materials

- A computer or tablet with access to the Internet
- If desired, scrap cardboard, copper tape, or aluminum foil and a Makey Makey

Process

1. Go to https://scratch.mit.edu/. If necessary, create an account. Click "Create" to get started on your code.
2. The Scratch coding screen is divided into four major sections.
   a. The "Stage" is located in the upper-right section of the screen. This is where your game or animation is acted out.
   b. Below this is the "Sprite area." This shows the characters, or sprites, that are used in your code. Clicking on each sprite allows you to control what they do and how they look. This area also allows you to add new sprites or delete ones you no longer want. You can also change the backdrops shown on the Stage.
   c. To the far left is the "Block Palette." It has three tabs: "Code," "Costumes," and "Sounds." Under the default "Code" tab, you can select various coding blocks to use in your program. Blocks are color coordinated and grouped by purpose. The "Costumes" tab has tools that let you edit or create sprites and backdrops. The "Sounds" tab let you select, edit, and create sounds to use in your program.
   d. In the center is the Scripts area, you will build the code for each sprite. When you select a sprite and then drag blocks into the Scripts area, you are telling that sprite what to do in the Stage.
3. In the "Sprites area," click on the cat. Remove him by clicking on the "X" in the upper right-hand corner.
4. At the bottom of the "Sprites" window, there is a "Choose a sprite" button. Click it, select the magnifying glass, and then find and select the robot sprite. Alternatively, campers can use the paintbrush to draw their own robot or use the import feature to import a graphic they have downloaded from the internet. Drag the robot to the center of the Stage.

5. Use the preceding method to add several drum sprites to the Stage. Drag each into a line of drums in front of your robot. If desired, use other sprites, create new ones, or download some from the internet (Figure 5.61).

FIGURE 5.61

6. Click on the first drum sprite. From the "Blocks Palette," select the yellow "Events" panel. Drag a "when key pressed" block to the Scripts area. There is a drop-down menu in the block, identified by a little downward arrow. As a default, it says "space." Click on the word "space," and select "left arrow" instead. This means that every time the left arrow is clicked, the code under this block will be activated.
7. Now we're going to make the drum sprite change the way it looks when the left arrow is clicked.
    a. From the "Blocks Palette," select the purple "Looks" panel. Drag the "next costume" block to the Scripts area, and move it until it clicks onto the "when key pressed" block. Each sprite can have several costumes. You can see them under the "Costumes" tab. If you selected drums from the Scratch library, your sprite will have at least two costumes. If you made your own, you will need to create additional costumes for your sprite, if you like. This is a fun way to make it look like your sprite is moving. The "next costume" block cycles through the costumes for the sprite in order from top to bottom and then back to the top again.
    b. From the "Blocks Palette," select the purple "Looks" panel. Drag the "change size by" block to the Scripts area, and move it until it clicks onto the "next costume" block. This will make the sprite change its size. The default is 10 percent. You can leave this or make it larger.
8. From the "Blocks Palette," select the fuchsia "Sounds" panel. Drag in the "play sound until done" block, and move it until it clicks onto the "change size by" block. Click on the drop-down menu, and select a sound. If you want to change or add a different sound, click on the "Sounds" tab at the top of the "Block Palette." On the bottom left, there is a "Choose a sound" button. Here you can select a sound from the library, import a sound you've downloaded, or create your own sound.
9. Now we need to undo the changes we made to the sprite's looks so that it is reset to the original for the next time the key is pressed. From the "Blocks Palette," select the purple "Looks" panel. Drag the "next costume" block to the Scripts area, and move it until it clicks onto the "play sound until done" block. Then drag the "change size by" block to the Scripts area, and move it until it clicks onto the second "next costume"

block. Change the number to be negative, or the opposite, of the size change you selected before by placing the minus sign in front of the number. Thus, if you had the drum change size by "10" before, it should change now by "–10."

10. Test your script by clicking on the yellow "when key pressed" block or by clicking on the left arrow.
11. Repeat steps 6–10 for each drum sprite. If you like, you can select the code from your first drum and use "CTRL+C" to copy it. Then you can go to the next drum and use "CRTL+P" to paste the code into the Scripts area. Be sure to change the code for each drum selecting different options from the drop-down menus for the "when key pressed" and "play sound until done" blocks.
12. Now we will code the robot sprite. Click on the robot sprite in the Sprite area. Let's "initialize" our robot and give it a starting place for the program.
    a. From the "Blocks Palette," select the yellow "Events" panel. Drag the "when [green flag] clicked" block to the Scripts area. Any code blocks attached to this block will take place immediately when the green flag in the upper-left corner of the Stage is clicked and the program starts.
    b. From the "Blocks Palette," select the blue "Motion" panel. Drag the "go to x y" block to the Scripts area, and move it until it clicks onto the "when [green flag] clicked" block. On the stage, click on the robot, and drag it to the place you would like it to begin the program. Now look in the Sprites area. There is a number next to "X" and a number next to "Y." These numbers refer to where a sprite is on the Stage. Enter those numbers into the white bubbles of your "go to x y" block.
    c. From the "Blocks Palette," select the blue "Motion" panel. Drag the "point in direction" block to the Scripts area and move it until it clicks onto the "go to x y" block. It should have the number 90 in the white bubble as a default. Leave that. This will make sure that the robot is standing up straight when you start.
13. Now let's make the robot dance.
    a. To make sure that the robot dances for the entire time the code is running, we'll need to use a loop to repeat code over and over. From the "Blocks Palette," select the orange "Control" panel. Drag the "forever" block to the Scripts area and move it until it clicks onto the "point in direction" block.
    b. From the blue "Motion" panel, drag over the "turn [right] degrees" block, and place it inside the "forever" block. The number in the white bubble should be 15. You can change this if you like.
    c. From the orange "Control" panel, drag over the "wait seconds" block, and place it under the "turn [right] degrees" block. The number in the white bubble should be 1. You can change this if you like; for example, you can set it to "0.5" to make the robot move faster.
    d. Repeat steps 13b and 13c, but instead use the "turn [left] degrees" block. At this point, if you test your robot, it should lean to the right and then back to the middle over and over.

e. Repeat steps 13b through 13d, but enter "-15" in the white bubbles. The robot should now tilt right, go back to center, tilt left, and then return to center. Adjust the time and degree of angle until your robot dances in a way you like.

14. It would be a lot of fun for the robot to move to each drum as it is played.

 a. From the yellow "Events" panel, drag the "when key pressed" block to the Scripts area. Change the drop-down menu so that it matches the key you selected for your first drum (for example, left arrow).

 b. From the blue "Motion" panel, select *either* the "go to position" block *or* the "glide to position" block. The "go to" block will have the robot instantly travel to the spot you selected. The "glide" block will have it move over time to that position. From the drop-down menu, select the drum that corresponds with your key press. If you are using the "glide" block, adjust the time until it looks the way you want, just as you did for the "turn" block earlier.

 c. After moving to the drum, you will need the robot to go back to the starting position. From the blue "Motion" panel, select the "go to x y" block or the "glide to x y" block. For X and Y, enter the numbers you selected in step 12b, your starting position.

 d. Repeat steps 14a through 14c for each of the other drums, changing the key press and position for each. You can either copy and paste as before or right-click on the code and select "Duplicate" from the pop-up menu.

15. If desired, use the blocks of the "Sound" panel to add sounds that the robot can make when different keys are pressed.

16. Click the green flag at the top of the Stage. Test your code. Can you play a different sound with each drum? Does the drum change size and costume and then return to its original look? Does the robot move to each drum like it should? Does the robot dance? Debug your code as needed. A sample copy of the completed code is provided (Figures 5.62 and 5.63).

For the Drums

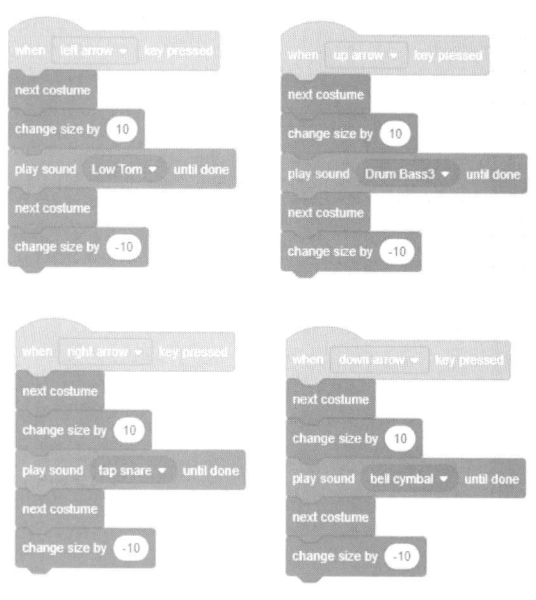

FIGURE 5.62

17. Once everyone has programmed their robots, it's time for the drum circle. The campers will "play" their robots together. The goal is to have the music played by each robot come together into a unified rhythm or song as a group. This means that you need to listen carefully to what others are playing and work to weave your drumbeats

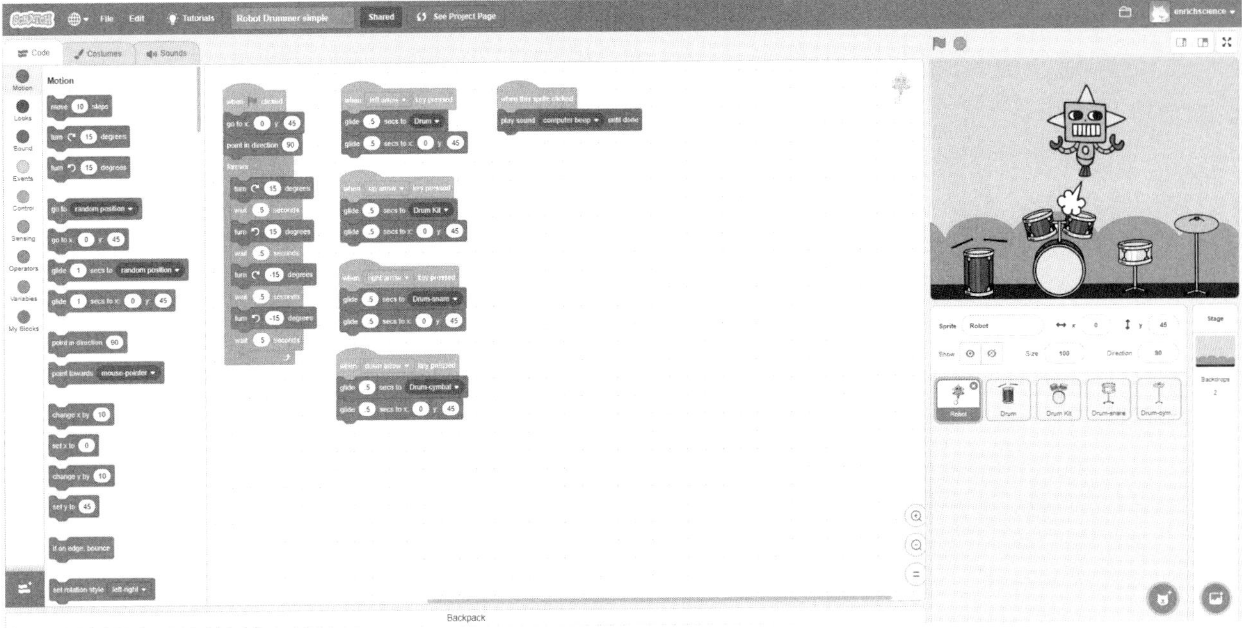

FIGURE 5.63

into the music of the group. It's like magic when everyone comes together and creates something completely new. Try it!

NOTE: Remix the code for this project at https://scratch.mit.edu/projects/279240141/

Take It Further

There are many excellent project guides and inspirations at the Makey Makey Labz website (https://labz.makeymakey.com/). You can also find many great projects in *20 Makey Makey Projects for the Evil Genius*, by Colleen and Aaron Graves (McGraw-Hill Education TAB, 2017).

PROJECT 9
Making Up a Maker Camp Song

COST: $

MAKE TIME: 15–30 minutes

If you ever attended a camp as a child, in all likelihood the camp had a camp song or chant that all the campers would enthusiastically sing while hiking or performing other activities. Why shouldn't Maker Camp have the same? It's not hard to do. Simply select a common and familiar song and modify the words to reflect your values. This is a great group brainstorming, get-to-know-you activity for the first day of camp. It tends to lead to a lot of giggles and good times.

Once you're done, you may want to have campers type up their creations so that every camper has a songbook to take home. Or get out the smartphones and record campers singing. (Do not share their videos unless you have permission from all parents or guardians.) Get out the green screen and create a video for your songs. Or use Scratch to code an animation based on your songs. The possibilities are endless.

Because every camp song will be personalized, below I provide some public-domain starting points. There are many lists online with similar songs, such as SongScouting (https://songscouting.wordpress.com/). A rhyming dictionary either online or in print is very helpful for this activity. Depending on their age and experience, some campers may need help breaking down songs by syllables, although you don't need to get too fussy about it.

Depending on your campers, you may need to set some guidelines about what language is appropriate and what is not. I usually ask campers to keep it "school appropriate" or "classroom appropriate." This often does the trick, but depending on your group and the culture of your specific camp, you may need to specifically restrict swear words, name calling, violent terms, references to weapons, religious references, and so on. Use your best discretion, and make this an opportunity for discussion about being inclusive.

"The Ants Go Marching"

(Parody of "When Johnny Comes Marching Home")

The ants go marching one by one, hurrah, hurrah
The ants go marching one by one, hurrah, hurrah
The ants go marching one by one,
The little one stops to suck his thumb
And they all go marching down to the ground
To get out of the rain.

The ants go marching two by two, hurrah, hurrah.
The ants go marching two by two, hurrah, hurrah.
The ants go marching two by two,
The little one stops to tie his shoe
And they all go marching down to the ground
To get out of the rain.

And so on for each number.

Campers could change the song to something like:

The campers are making one by one, hurrah, hurrah
The campers are making one by one, hurrah, hurrah
The campers are making one by one,
The robot stops to eat bacon
And they all go making round and round
Until the end of camp.

"This Old Man" is a similar repetitive rhyming song that you could use. The ever popular "Boom Chicka Boom" offers a lot of opportunity for customization and rhyming as well.

Repeat each line after the song leader:

I said a Boom Chicka Boom
I said a Boom Chicka Boom
I said a Boom Chicka Rocka Chicka Rocka Chicka Boom
Uh huh
Oh yeah
One more time [job/location/item] style.

Maker Style
I said a new something new
I said a new something new
I said a new make a, break a, fix a, build a, invent a something new.

"Rose, Rose" is a traditional song from the 1400s that is typically sung in the round:

Rose, Rose, Rose, Rose,
Shall I ever see thee wed?
I will marry at thy will, sire,
At thy will.

Campers could change it into a supportive song for coding, such as:

Code, code, code, code,
Will I ever see it run?
I will debug all the errors.
Fixing code.

Similar rounds include "Row, Row Your Boat," "Three Blind Mice," and "White Coral Bells." Why not break campers into groups of three or four, give each group a different starter song, and see how many new rounds they can create?

For campers who may need a little more structure, try a Mad Libs format, and give them blanks they need to fill in. For example:

On top of ________________,
3 syllables, place

All covered in ________________,
1 syllable, noun

I lost my ________________ ________________,
1 syllable, adjective 2 syllables, noun

By ________________ too ________________.
2 syllables, verb 1 syllable, adverb

This could become a sad tune about 3D printing:

On top of Makerbot,
All covered in dust,
I lost my new Yoda
By heating too low.

Of course, some liberties can be taken with the rhythm. Here's a cautionary tale about electronics:

On top of my robot,
All covered in sparks,
I lost my best micro:bit
By p'wering 6 volts.

PROJECT 10
Mixing Music with the NeoTrellis

COST: $$

MAKE TIME: 1–3 hours

The Adafruit NeoTrellis M4 is an all-in-one Musical Instrument Digital Interface (MIDI) audio board. It has a grid of 4 × 8 neopixels (special color-changing LED lights) that act as input. It literally puts sound at the touch of a button. The NeoTrellis can be used as anything from a sound effects generator to a synthesizer.

It can be controlled with CircuitPython and Arduino, so the coding possibilities are pretty endless. Priced at around $60 and ready to use out of the box, the NeoTrellis has quickly developed a devoted fan base among computer nerds and music lovers alike.

If you have campers who want to take their music making to whole new levels, the NeoTrellis may be the perfect technology. Even if your campers aren't experienced coders, they can download files coded by others and remix them, adding their own sound file libraries or simply mixing what is already available.

Of course, for those who are interested, jumping into learning Python is a great idea. It's one of the hottest languages out there, and it isn't difficult to learn. Adafruit has done a great job documenting its products on its website and providing lots of tutorials and open-source projects with which campers can experiment.

In this section, I present three starter projects and ideas for expansion and experimentation. If you are more interested in playing with the NeoTrellis and using it as a creative tool rather than working on coding or sound editing, that's fine. The code and sound files are all available for download on GitHub at https://github.com/KaleidoscopeSci/NeoTrellis.

Materials

- Adafruit NeoTrellis M4 with Enclosure and Buttons Kit Pack
- USB data cable
- USB-powered speakers
- A computer with USB ports and an internet connection
- Sound editing software such as Audacity
- Python editing software such as Mu
- A USB microphone (optional, for creating your own sounds)

The Sound Effects Board

This is the simplest of the projects. First, you'll need to select sound files for your project and make sure that they are edited properly for the NeoTrellis. Then you'll download the CircuitPython code and edit it. Because you're hacking premade code, you don't need to be an ace programmer to make this work, trust me (Figure 5.64).

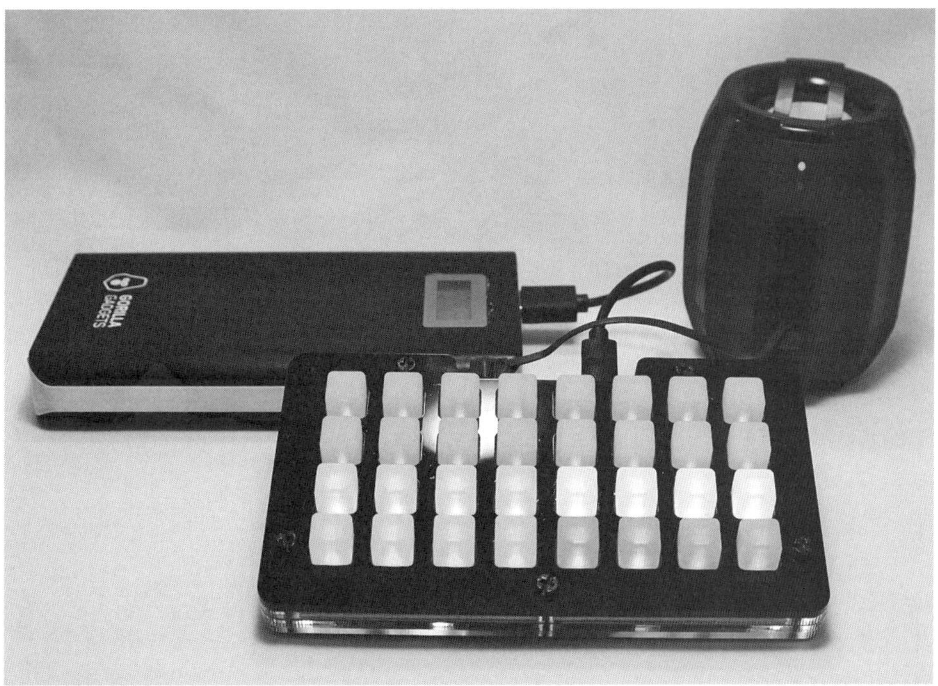

FIGURE 5.64

Process: Preparing the Sounds

All the sound files for the NeoTrellis must be PCM 16-bit Mono WAV files at a 22-kilohertz sample rate. Okay, that was probably scary to read. Don't worry. *PCM* means "pulse-code modulation." It's the standard form of digital audio in computers. *Mono* refers to the number of channels used: *mono* is one, whereas *stereo* is two. *WAV* refers to a specific file type (like MP3). WAV files are uncompressed and binary. Because of their nature, you can reduce the file size while keeping sound quality. This is important for us.

The *16-bit* refers to the bit depth, which is the number of bits of information in each sample within the sound. To save space, digital music samples sounds at regular intervals rather than saving the complete, continuous wave of an analog sound. The more bits it stores for each sample, the more levels, or *depth*, the sound will have.

The *22-kilohertz* refers to the sample rate. This is the number of times the audio is measured, or sampled, per second. The more frequently it is sampled, the richer and of better quality is the sound. CDs are usually 44 kilohertz.

So we're saving our sound files with these setting to save space. Sound tends to make very large digital files. To fit 32 files on our board, we need to make sure that they take up as little space as possible without sacrificing too much quality. For this reason, you'll want to keep your sound file fairly short as well. I usually don't go over 5 seconds.

Once you've found or created your sound files, you'll need software to edit them. Audacity is free and open source. You can use it to both create and edit your files. Specific details on using the software as well as download instructions can be found at https://learn.adafruit.com/microcontroller-compatible-audio-file-conversion/check-your-files (Figure 5.65).

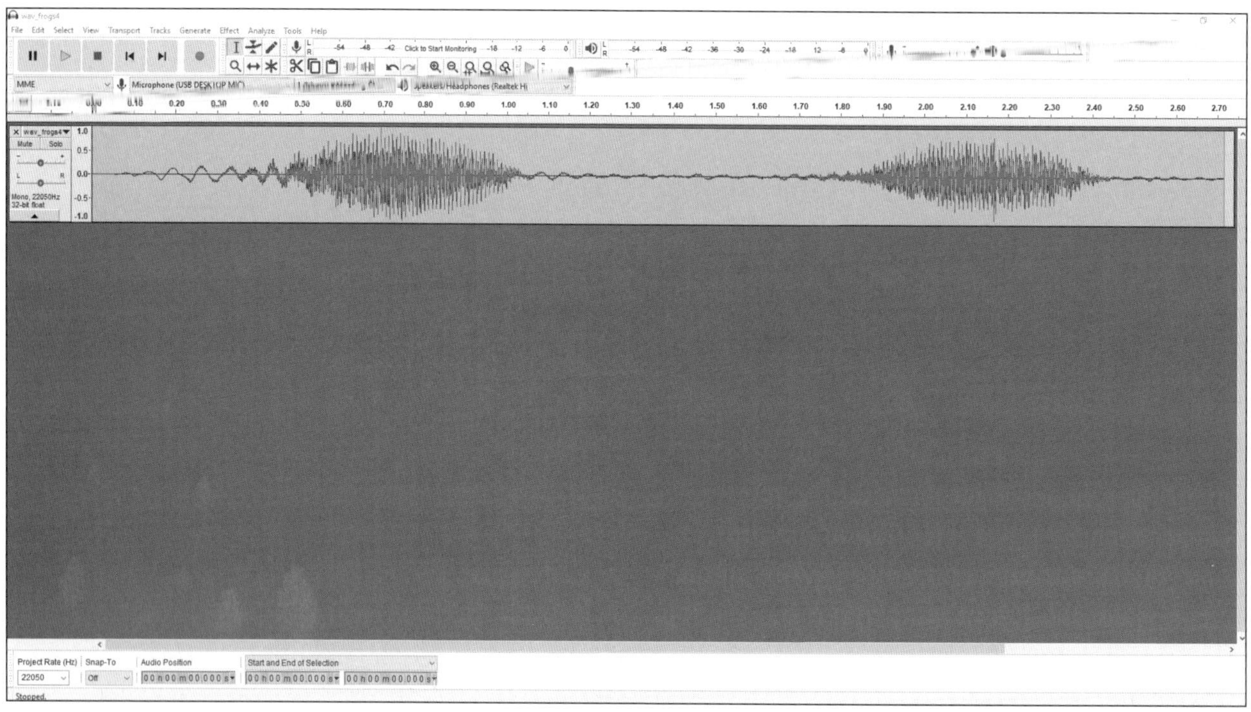

FIGURE 5.65

Give your files simple names, and put them all together into one folder.

If this feels like a lot of work, never fear. It can be time consuming, but it's not difficult. This means that it's the perfect project for campers to divide up and conquer.

If you prefer to download files already prepared for the NeoTrellis, you have lots of options. For example, I have a collection of nature sounds from the National Park Service that you can download to bring the great outdoors inside. It's available on my GitHub page at https://github.com/KaleidoscopeSci/NeoTrellis. A quick search on Adafruit (http://www.adfruit.com) for NeoTrellis projects will also yield several options.

Process: Preparing the NeoTrellis

Before you can use your NeoTrellis, you need to get it ready to accept your code. Once you have everything set up, your NeoTrellis will work like a drive on your computer, similar to when you use a USB memory stick or SD card for photos. This makes it easy to just drag and drop files on the device. And your code editing will automatically update changes to the device when you save. So, while it takes a little while to get things set up, it's not difficult, and it makes using the NeoTrellis a snap. You'll need to

- Assemble the board and enclosure. Instructions are available at https://learn.adafruit.com/adafruit-neotrellis-m4/assembly.
- Install CircuitPython on the NeoTrellis. Instructions and all the downloadable files are available at https://learn.adafruit.com/adafruit-neotrellis-m4/circuitpython.

- Upload "Libraries" to the NeoTrellis. These are files with information the NeoTrellis needs to perform various actions. You basically select the ones you need, and then drag and drop them onto the NeoTrellis. Instructions are available at https://learn.adafruit.com/abc-soundboards-with-neotrellis/setting-up-circuitpython.
- Download and install software to edit CircuitPython, such as the free Mu editor. Learn more about this process at https://learn.adafruit.com/welcome-to-circuitpython/installing-mu-editor.

Okay. At this point you should have your NeoTrellis ready to go.

Process: Creating the Sounds of Nature

Let's start by getting our sound files onto the NeoTrellis. Select the folder where you have stored your files. Drag it into the main directory of the CIRCUITPY drive.

Now go to the GitHub link and download the file code_nature.py. Copy the file, and rename it code.py. Then drag it onto the CIRCUITPY drive. Open Mu (or your chosen Python editor), and select "Mode." Your NeoTrellis should show up as a choice. Select it. Click on "Load," find the CIRCUITPY drive, and select the code.py file. Now you can make changes, and when you click "Save," the changes will be automatically update on the NeoTrellis.

For example, you may want to change the colors of the buttons. That's easy to do. Each color is coded by a hexadecimal number preceded by "0x." You can code a simple name (such as "RED") to stand in for a hex color (such as 0xFF0000), as I do under "Custom colors for keys." You can also add any colors you like. Many reference charts for the hex codes for various colors are available online, such as this one from Wikipedia: https://upload.wikimedia.org/wikipedia/commons/9/95/Xterm_color_chart.png.

You may want the sounds in a different order on your board. Or perhaps you're using different sound files completely. Make sure that the "SAMPLE_FOLDER" points to the folder name with your sound files (the one you loaded onto the CIRCUITPY drive). The code section that lists the "SAMPLES" includes the name of the WAV file and the color you'd like for each button. Files will be ordered starting in the upper-left corner (when the board is oriented with the power and sound ports at the bottom) and moving horizontally across the board. Arrange your files as you prefer (Figure 5.66).

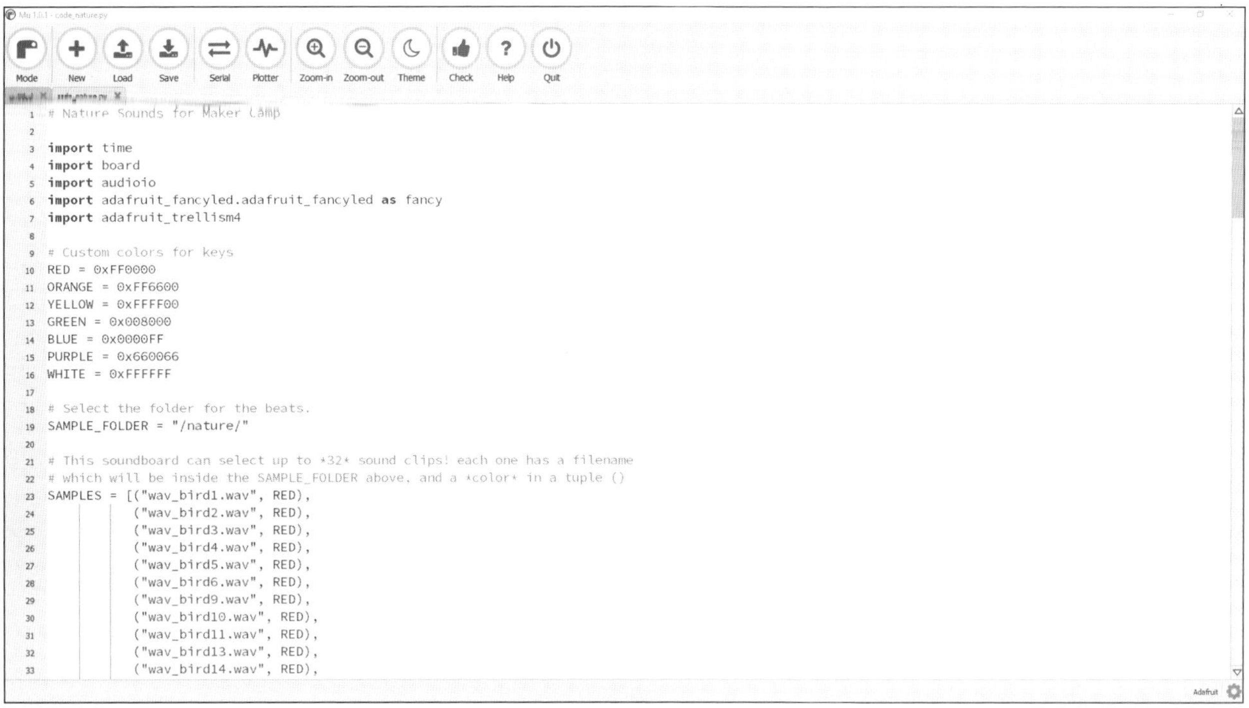

```
# Nature Sounds for Maker Camp

import time
import board
import audioio
import adafruit_fancyled.adafruit_fancyled as fancy
import adafruit_trellism4

# Custom colors for keys
RED = 0xFF0000
ORANGE = 0xFF6600
YELLOW = 0xFFFF00
GREEN = 0x008000
BLUE = 0x0000FF
PURPLE = 0x660066
WHITE = 0xFFFFFF

# Select the folder for the beats.
SAMPLE_FOLDER = "/nature/"

# This soundboard can select up to *32* sound clips! each one has a filename
# which will be inside the SAMPLE_FOLDER above, and a *color* in a tuple ()
SAMPLES = [("wav_bird1.wav", RED),
           ("wav_bird2.wav", RED),
           ("wav_bird3.wav", RED),
           ("wav_bird4.wav", RED),
           ("wav_bird5.wav", RED),
           ("wav_bird6.wav", RED),
           ("wav_bird9.wav", RED),
           ("wav_bird10.wav", RED),
           ("wav_bird11.wav", RED),
           ("wav_bird13.wav", RED),
           ("wav_bird14.wav", RED),
```

FIGURE 5.66

Click "Save." The NeoTrellis should restart, and you can use your program. Try pressing the buttons and checking the sound files. If the board goes black and does not light up, this means that you have a bug in your code. Carefully check that you haven't accidently made a typo. If all else fails, go back to your backup copy and start again.

The Sound Mixer

With a sound mixer, you can play multiple WAV files at once and have them repeat over and over to make new patterns. With the nature sounds I've provided, you can create the illusion of crickets chirping at night or birds singing by a stream. Upload human beat box sounds or drum beats, and you have another fun application.

Depending on what you want to do, you will want to download the "8-Step Drummer," by John Parks (available at https://learn.adafruit.com/trellis-m4-beat-sequencer/eight-step-simple-sequencer), which lets you use only four sounds but let's you create eight beat patterns. This is great for mixing drum beats. My own code_naturemixer.py uses eight sounds for a four-beat pattern at a much slower tempo, perfect for mixing natural sounds.

Both programs use the NeoTrellis' accelerometer to allow you to adjust the tempo of the pattern. Tilt the board to the right to speed it up and to the left to slow it down.

Whatever you choose, the process is the same:

1. Download the Python code of your choice. Make a copy, and rename it code.py.
2. Drag and drop code.py over to the CIRCUITPY drive.
3. Open Mu, and connect the NeoTrellis.
4. Edit the code for the color and sound files of your choice.
5. Click "Save" to update the code on the drive.
6. Test the results.

Sixteen-Step Sequencer and Sample

Also by John Parks, the "16-Step Sequencer" gives your campers the creative space to mix eight sounds with eight additional sounds they can record on the spot to the board. This code also offers up great tools such as low- and high-pass filters, volume control, tempo control, repeats, muting, and jumps. Campers will feel like they landed in a *Pitch Perfect* sequel.

All the documentation, downloads, and instructions are available at https://learn.adafruit.com/trellis-m4-beat-sequencer/sixteen-step-sample-seqeuncer.

You will need to set your board back to the original Arduino setup. Basically, this requires resetting the device so that it shows up as TRELSM4BOOT. Then you drag a new firmware file onto the device. There are complete instructions on the way to do this on the website above. It's pretty painless.

If you don't have a NeoTrellis but still want campers to experiment with sound sequencing, try my Scratch version at https://scratch.mit.edu/projects/282414002/ (Figure 5.67).

Take It Further

There are many other projects for NeoTrellis available on the Adafruit website. What do you think your campers will enjoy? With new ideas being uploaded every day, there are many ways to use this device. And if you have students interested in learning Python or Arduino, there are books to get them started on coding their own creations for the NeoTrellis. Check out these books from your local library:

- *Python for Kids: A Playful Introduction to Programming*, by Jason Briggs (No Starch Press, 2012)

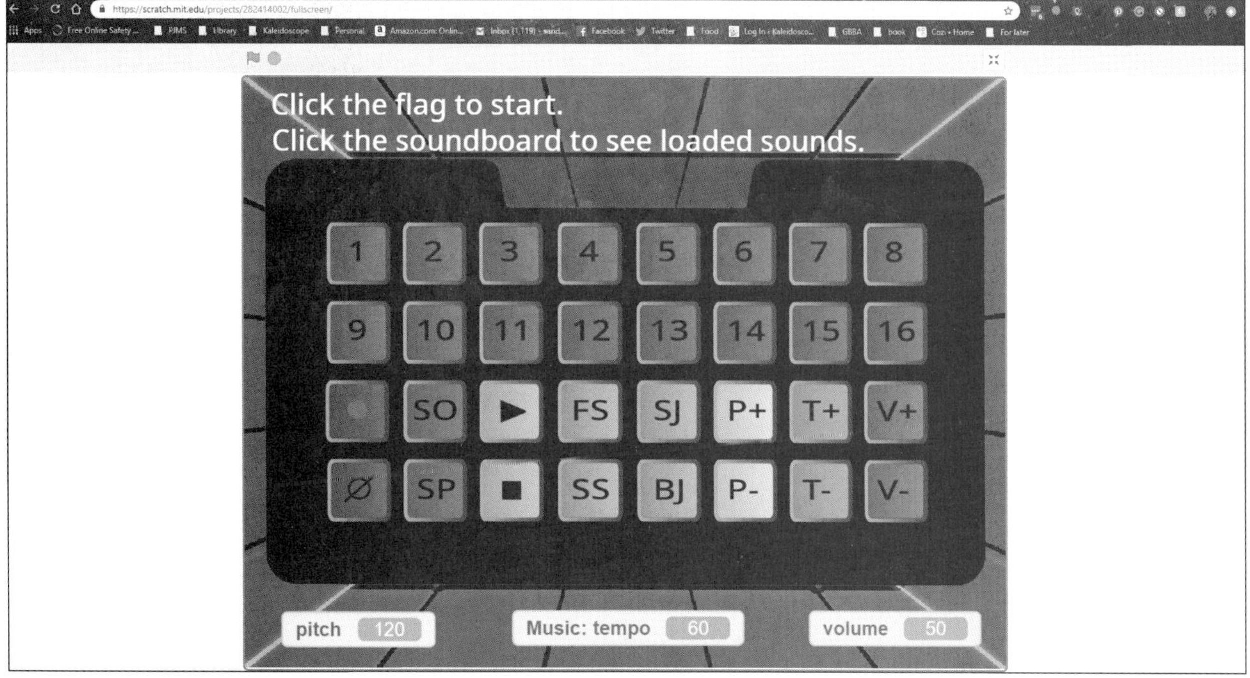

FIGURE 5.67

- *Python for Microcontrollers: Getting Started with MicroPython*, by Donald Norris (McGraw-Hill Education TAB, 2016)
- *Programming Arduino, Second Edition*, by Simon Monk (McGraw-Hill Education TAB, 2016)
- *30 Arduino Projects for the Evil Genius, Second Edition*, by Simon Monk (McGraw-Hill Education TAB, 2013)

CHAPTER 6

Camp Food

Throughout the year at my Maker Clubs, I periodically hold "kitchen chemistry" classes because the amazing science behind our food is perpetually fascinating to kids. Cooking and baking provide early insights into chemistry, physics, and biology, so they should be standards in any camp or classroom. These activities are also fundamental making experiences, where traditional knowledge, technical skills, and creativity blend together, thus creating delicious new discoveries. The maker movement has combined this with technology, resulting in pancake bots and 3D-printed sugar candies, as makers push the boundaries far beyond anything Grandma used to make.

While any of these activities will be a great addition to your Maker Camp program, please consider allergens in your planning process. Make sure that you are aware of any allergy issues that your campers may have. Ensure that your projects and materials are allergen-free if necessary. Have appropriate first aid on hard, just in case. You'll also want to check to see if you have vegan or vegetarian campers because this also can affect your recipes. Think of mitigating food allergies and preferences as teaching moments and opportunities to try something new.

These projects tend to be some of the messiest camp activities. Plastic drop cloths and disposable gloves are a must for many projects. Be sure to discuss the risks of contaminating food with germs and emphasize procedures that will reduce any chances of passing along the critters. You'll also want to have bags or boxes available for campers to take home their creations.

PROJECT 1
Making Marshmallows

COST: $

MAKE TIME: 30 minutes

Marshmallows are not a new food. The ancient Egyptians made a similar treat using the root of the marsh mallow, a common swamp flower with a carrot-like root. The roots of these plants contain a substance called *mucilage*, which is a thick, gluey substance that helps with food storage and seed germination. When the root was collected, dried, and ground, it can be used in a variety of ways. When added to hot water as a tea, it is used to treat sore throats and digestive issues. When added to warm honey and whipped, you get a sweet treat or a liquid bandage for injured skin—your choice!

These days, we don't use the marsh mallow to make treats anymore; it's more likely to be included in glue. However, the name has stuck for the fluffy confection. Modern marshmallows are made with sugar, corn syrup, and gelatin to replace the mucilage. Gelatin is made from the collagen of animals and has the ability to unravel protein strings when heated and then coil them again when cooled, trapping liquids in a net and forming a gel. This makes it very useful in the creation of many types of desserts.

Making your own marshmallows isn't very difficult, and the results are usually fantastic. They are tastier and have better texture than store bought marshmallows. Plus, you can add color and flavor to customize your treat. Making your own marshmallows for s'mores may take some extra time and planning, but it's worth it!

Because you are working with heat, take fire safety precautions. Work on heat-safe surfaces and have a fire extinguisher nearby. You may also want to have campers wear goggles when you mix the ingredients.

Materials

- 3 packages unflavored gelatin
- 1½ cups granulated sugar
- 1 cup light corn syrup
- 1 cup water
- ¼ teaspoon kosher salt
- 1 tablespoon pure vanilla extract or other flavoring
- Food color if desired
- Confectioners' sugar, for dusting
- Measuring spoons and cups
- Hot plate or stove
- Heat-safe spoon
- Saucepan
- Candy thermometer
- Stand mixer with whisk attachment or a hand mixer and large bowl
- 8- × 12-inch nonmetal baking dish

Process

1. Combine the gelatin and ½ cup of cold water in the bowl of an electric mixer, and allow the mixture to sit. Attach the whisk to the mixer.
2. Combine the sugar, corn syrup, salt, and ½ cup water in a small saucepan, and cook over medium heat until the sugar dissolves.
3. Raise the heat to high. Cook the mixture until it becomes a syrup, at about 240 degrees F on the candy thermometer. Remove from the heat (Figure 6.1).

FIGURE 6.1

4. With the mixer on low speed, slowly pour the sugar syrup into the dissolved gelatin.
5. Put the mixer on high speed, and whip the mixture until it is very thick, about 15 minutes (Figure 6.2).

FIGURE 6.2

6. Add the vanilla or any desired flavoring and colors. Mix thoroughly.
7. Dust your dish with confectioners' sugar. Pour the marshmallow mixture into the pan, smooth the top, and dust with more confectioners' sugar (Figures 6.3 and 6.4).

FIGURE 6.3

FIGURE 6.4

8. Allow the marshmallow to stand uncovered overnight until it is dry to the touch.
9. Turn the marshmallow onto a board, and cut it into small squares (or use cookie cutters dipped in powdered sugar). Dust the small bits with more confectioners' sugar (Figure 6.5).
10. Store them covered in plastic wrap.

FIGURE 6.5

Take It Further

Marshmallows can be decorated with food-safe markers or with watered-down food coloring and paintbrushes. Try flavoring your marshmallows. For example, adding orange extract with the vanilla and a little orange food dye yields "creamsicle" marshmallows.

PROJECT 2
Making Chocolate

Making chocolate is one of the most consistently loved activities in my Maker Camps. There are so many possibilities, so it's not a surprise that it's a perpetual favorite. The fact that you get to enjoy your tasty treats after the hard work is done may have something to do with this activity's popularity as well.

When making chocolate, you have a lot of options for your base materials. You can pick up chocolate chips at the grocery store or candy melts at the craft store or get hands-on premium chocolates. Any of these will work. I find candy melts to be the easiest to use with large groups while also being cost effective (use those coupons as teacher discounts!). Candy melts also come in lots of color and flavors, which make them lots of fun for campers who want to get creative.

As for basic equipment, the most important consideration is how you will melt your chocolate. If you have access to a microwave, that is likely your easiest option. Just put your chocolate in a microwave-safe container, and "nuke" it in 10-second increments, stirring it frequently.

If you only have access to a stovetop or hot plate, plan to use a double-boiler. This means that you'll have one pot heating water below a second, well-fitted pot in which you will melt the chocolate. You'll also need a candy thermometer. Milk chocolate should not get above 105 degrees F. This is not the easiest setup, but it gets the job done. Take appropriate precautions with all the heated elements. For a safer option, try fitting a pan inside a slow cooker filled with water.

If, like me, you suspect that you may be making chocolate frequently, use those coupons and invest in a standalone electric chocolate melting pot. I have one with several silicone pot inserts, and it saves a lot of headaches.

If you are planning to use traditional chocolate rather then candy melts, you will want to *temper* the chocolate. The tempering process realigns the sugar crystals in chocolate, stabilizing its structure and making it shiny and smooth when finished. It also helps to keep chocolates from melting in your hands. Basically, you allow the chocolate to cool to 82 degrees F and then reheat it to between 85 and 87 degrees F before use. Alternatively, you can melt two-thirds of your chocolate and remove it from heat. Then drop in the remaining third and stir slowly to temper the chocolate. I tend to use the second method.

You also want to have a way to chill the chocolate. Otherwise, it may take quite a long time to set. Obviously, the best option is a refrigerator or freezer. If you don't have access to either, a cooler with ice will work reasonably well, although it doesn't give you much space. If chiller space is an issue, choose smaller chocolate molds that will allow the chocolate to harden more quickly.

Next I detail several chocolate-making methods, from the simplest to the most challenging. If desired, it can be a lot of fun to offer colored sugars, sprinkles, and crushed candies to add to the chocolates you create. Have fun and get creative!

Using Commercial Molds

COST: $–$$

MAKE TIME: 15 minutes, plus time to harden the chocolate

This is probably one of the easiest ways to introduce chocolate making to your campers. Plastic chocolate molds are inexpensive and available in a wide variety of designs. There are molds for shaped chocolates, filled chocolates, lollipops, pretzel log toppers, and more. You may even be able to find molds at local thrift stores.

Many commercial molds allow you to paint the mold before pouring in the chocolate, creating colored pieces. To do this, melt candy melts in different colors, and place some in a metal muffin tin to create a pallet. Place the muffin tin on a heating pad or electric griddle to keep the chocolate melted. Use small paint brushes to add color to the molds.

Materials

- Plastic chocolate molds
- Bulk chocolate or candy melts
- A large spoon or a small ladle
- A method to melt the chocolate (see above)
- A method to chill the chocolate (see above)
- Any items you plan to use for decorations (see above)
- A cutting board or cookie sheet

Process

1. Melt and temper your chocolate.
2. Add your decorative elements.
3. Ladle melted chocolate or candy melts into the molds, taking care not to go above the lip of the mold.
4. Gently tap the mold to remove any air bubbles.
5. Chill the chocolates until solid (10–20 minutes in the refrigerator).
6. Unmold the chocolates by inverting them and gently smacking the mold against the cutting board or cookie sheet. If the chocolates don't release, let them warm up a bit, or place the bottom of the mold in warm water. Then try inverting again.

Using Cookie Cutters as Molds

COST: $–$$

MAKE TIME: 15 minutes, plus time to harden the chocolate

This is one of the easiest ways to mold chocolates. If you do the 3D-printer cookie-cutter later in this chapter, campers can use their own creations to make chocolates too! Of course, even if you didn't make your own cookie-cutter marshmallows, many different options are available in stores. I do find that metal works a little better than plastic because it releases better, but either way, this is a simple and fun way to make your own chocolate bars or pieces. Oh, and it works for fudge too!

Materials

- Cookie cutters (rounded shapes work best)
- Chocolate or candy melts
- A large spoon or a small ladle
- A method to melt the chocolate (see above)
- A method to chill the chocolates (see above)
- Any items you plan to use for decorations (see above)
- A metal cookie sheet
- Parchment or wax paper

Process

1. Melt and temper your chocolate.
2. Chill your cookie sheet in the freezer for 10–15 minutes. Line it with parchment paper.
3. Place your cookie cutters on the cookie sheet.
4. Slowly ladle chocolate or candy melts into each cookie cutter (Figure 6.6).
5. Add any decorative elements.
6. Chill the chocolates until solid (10–20 minutes in the refrigerator).
7. Unmold the chocolates gently by smacking the cookie cutter against the cookie sheet. If the chocolates don't release, let them warm up a bit, and then try again. If needed, use a knife inserted between the chocolate and the cookie cutter (Figure 6.7).

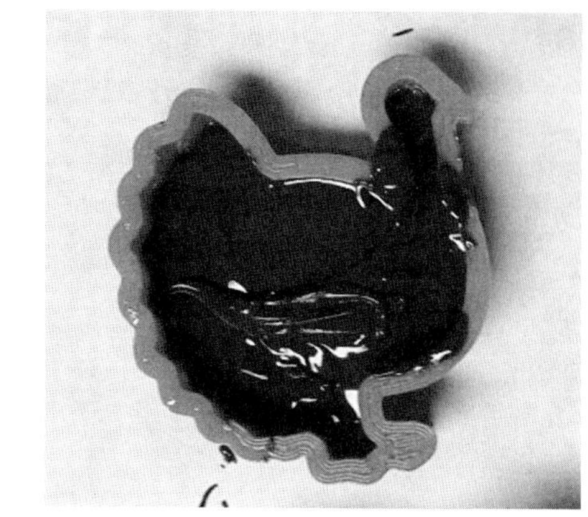

FIGURE 6.6

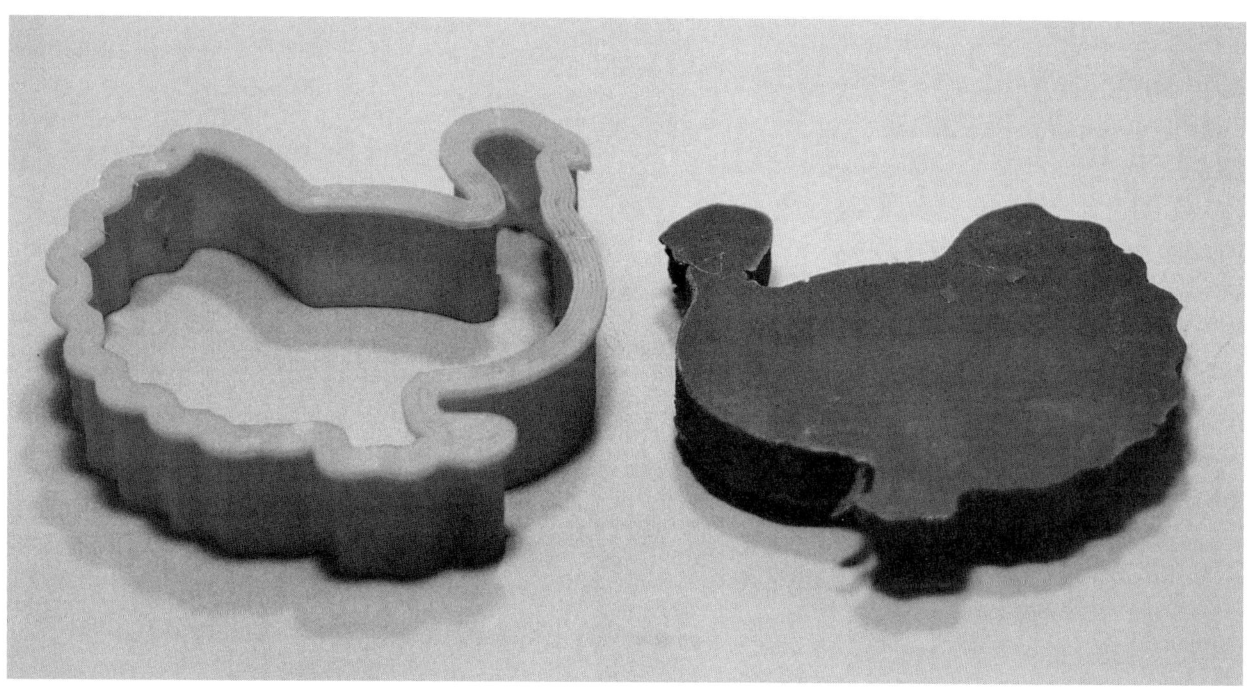

FIGURE 6.7

Brown Sugar Chocolate Molds

COST: $–$$

MAKE TIME: 15 minutes, plus time to harden the chocolate

There is something really satisfying about molding chocolate in sugar. This method is a lot like making sand candles, if you are familiar with the process. The fun thing here is that you can use just about any item to create your mold. Try Legos, action figures, brooches, or rubber stamps (cleaned well, of course). Those magnetic letters that every kid has at home? Wash them and use them! Get creative. By the way, the sugar is okay to use for baking afterward, so that's a bonus. As a note, your chocolates will have a grainy texture on the surface, but that adds to the charm.

Materials

- Items to press, washed and dried
- Chocolate or candy melts
- A large spoon or small ladle
- A method to melt the chocolate (see above)
- A method to chill the chocolates (see above)
- Any items you plan to use for decorations (see above)
- A metal cookie sheet or roasting pan
- 2-pound bag of brown sugar

Process

1. Pour the brown sugar onto the cookie sheet or roasting pan. Pack the sugar to make a dense coating, at least 1 inch thick.
2. Press your chosen object into the sugar, taking care not to go all the way to the bottom of the sheet or pan. Do not twist or turn the object. Press it straight in and then pull out. Be gentle (Figure 6.8).

FIGURE 6.8

3. Melt and temper your chocolate.
4. Slowly ladle the melted chocolate or candy melts into the brown sugar impression (Figure 6.9).
5. Add any decorative elements.
6. Chill the chocolates until solid (10–20 minutes in the refrigerator).

FIGURE 6.9

7. Pull the molded chocolates out of the sugar. If desired, use a brush to remove excess sugar, or briefly run the chocolate under cold water and then dry it (Figure 6.10).

Creating Silicone Molds

COST: $$

MAKE TIME: 30–45 minutes, plus time to harden the chocolate

While it may be tempting to use the same method outlined in Chapter 4 for creating silicone molds to use with chocolate, that process is not food safe at all. But all is not lost! There are commercial products available to create food-safe silicone molds for use with chocolate. Smooth-Sil and EasyMold Silicone Putty are popular options. However, my favorite is ComposiMold. That company offers a variety of mold-making options, including a reusable food-safe compound that you just need to heat in a microwave to use. So easy!

FIGURE 6.10

Follow the directions for your specific product to make a mold of your item. Then follow the same melting, tempering, pouring, and chilling procedure outlined earlier for the cookie-cutter method and brown sugar method. Store-bought molding materials are more expensive than the other methods described earlier, but they provide a level of detail that you can't find otherwise, and the molds may be reusable many times over.

3D-Printed Chocolate Molds

COST: $$$

MAKE TIME: 15–30 minutes to design, time varies to print

Unless you have heated vacuum formers available, the easiest way to create a custom chocolate mold is to use 3D-printing technology. Options are somewhat limited with free apps such as Tinkercad, but this project will give campers the opportunity to experiment.

If you have access to a flexible filament, use it. You will be able to cast more complex and detailed pieces using a material that can flex and release your hardened chocolate without sticking or cracking. You can also print with polylactic acid (PLA), but you must keep the design simple, without a lot of detail or sharp angles. With PLA, seal the print with a water-based acrylic spray sealant, and allow it to dry completely. This will help smooth any layers, making it easier to release the chocolate. Chill the mold before use, and use a cooking spray before pouring the chocolate to help with the release. Do not fill the mold all the way to the lip. Leaving a little space makes it easier to remove the chocolate later.

Always look for filaments that are considered food safe. If you are unsure, contact the seller or manufacturer. Do not use acrylonitrile butadiene styrene (ABS). It is not considered food safe.

Keep in mind that because there can be tiny gaps between layers of 3D prints, they may allow food particles to get caught in those gaps. Such areas can become breeding grounds for bacteria Don't expect to use your mold over a long period of time because it may become contaminated and unsafe. Also consider that brass hot ends may contain traces of lead, so check your equipment.

Even with these concerns, it is still considered safe to use 3D-printed molds because the chocolate will be in contact with the mold for such a short period of time. If you are concerned, campers can design items for 3D printing and then create molds from them using the brown sugar or silicone mold methods outlined earlier.

Materials

- CAD software, such as Tinkercad
- PLA filament, food safe
- 3D printer of your choice
- Chocolate or candy melts
- A large spoon or small ladle
- A method to melt the chocolate (see above)
- A method to chill the chocolates (see above)
- Any items you plan to use for decorations (see above)

Process

1. Go to tinkercad.com. Create an account, if necessary. Click on "Create."
2. From the menu on the right, select and drag a "box" or "column" onto the work plane. This will be the base of your chocolate mold. Click on a corner of the box. Small white boxes will display. Use them to drag the shape to a size of 50 × 50 millimeters. Click on the item again, and use the white

box on the top of the image to make the depth of object 15 millimeters (Figure 6.11).

3. Select and drag "Half sphere" onto the work plane. Use the white boxes to set the height and width to 35 millimeters. Set the depth to 13 millimeters (Figure 6.12).

4. Rather than selecting a color in the "Shape" toolbox, set the object to "Hole." This will allow you to use the shape to cut out a section from the box.

5. Click on the "Flip" icon, shown as two triangles in

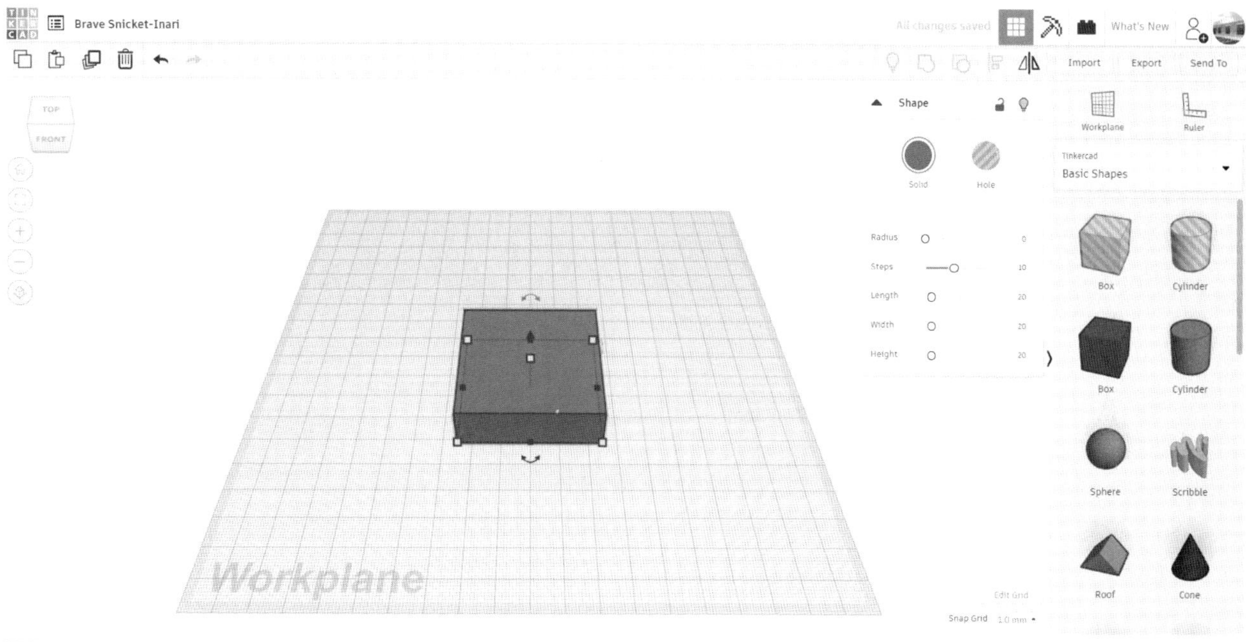

FIGURE 6.11

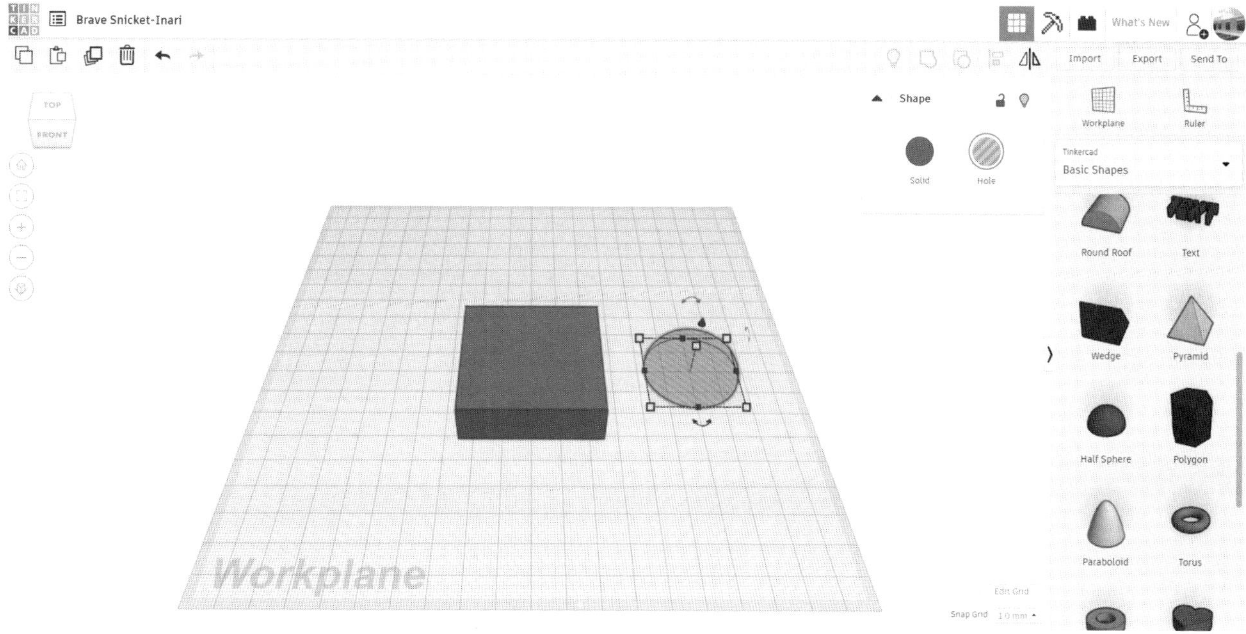

FIGURE 6.12

the upper right-hand corner. Flip the object so that the wide circular base is facing upward and the curve is facing the work plane.

6. Select the "Half sphere." Select the black cone above the object. Drag the cone so that the "Half sphere" is 2 millimeters above the work plane.
7. Drag the "Half sphere" over to the box. Center it on the box. If needed, select both objects, and use the "Align" tools (which look like two bars on a graph in the upper right-hand corner) to make sure that the "Half sphere" is centered. Group the objects. This is the base of the chocolate mold (Figure 6.13).
8. Select a shape to emboss onto the chocolate. Select a simple shape without a lot of details, sharp angles, or straight sides.
9. Resize the embossing object so that it is roughly 15 millimeters wide and 5 millimeters deep. This will vary for each shape, depending on its contours, so be prepared to adjust as needed (Figure 6.14).
10. Move the embossing object to the base. Using the "Align" tools, center the object. Group the items (Figure 6.15).
11. Export the object and 3D-print it.
12. Melt and temper your chocolate.
13. Ladle melted chocolate or candy melts into the molds, taking care not to go above the lip of the mold.

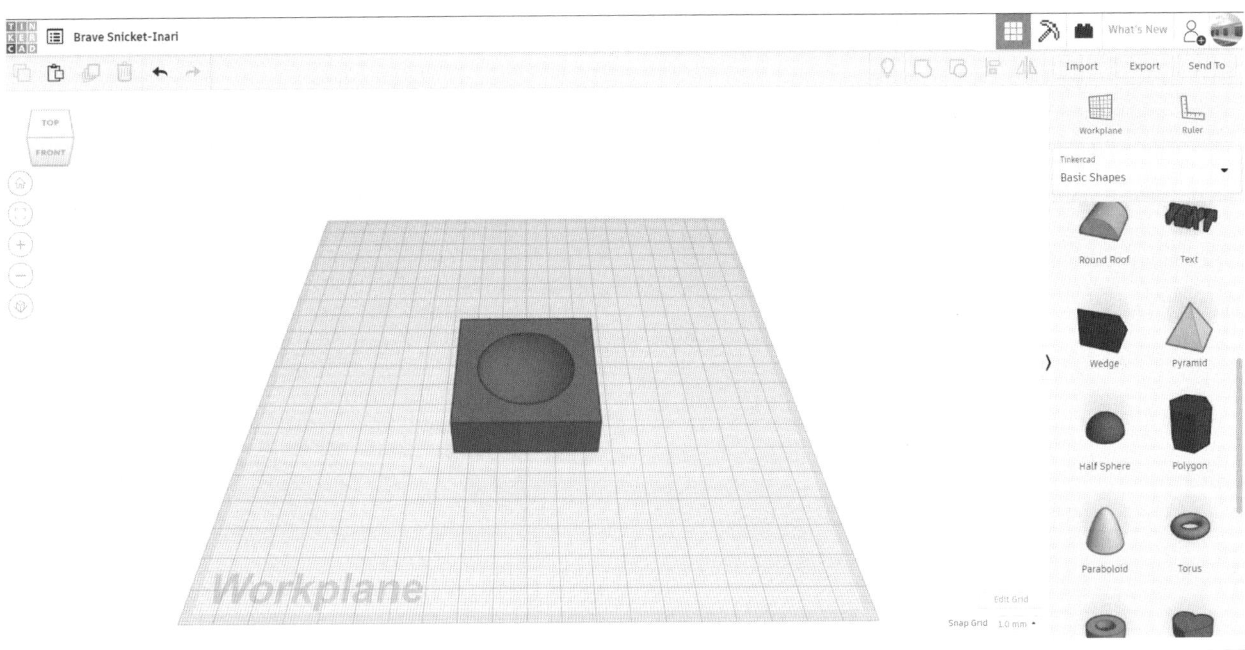

FIGURE 6.13

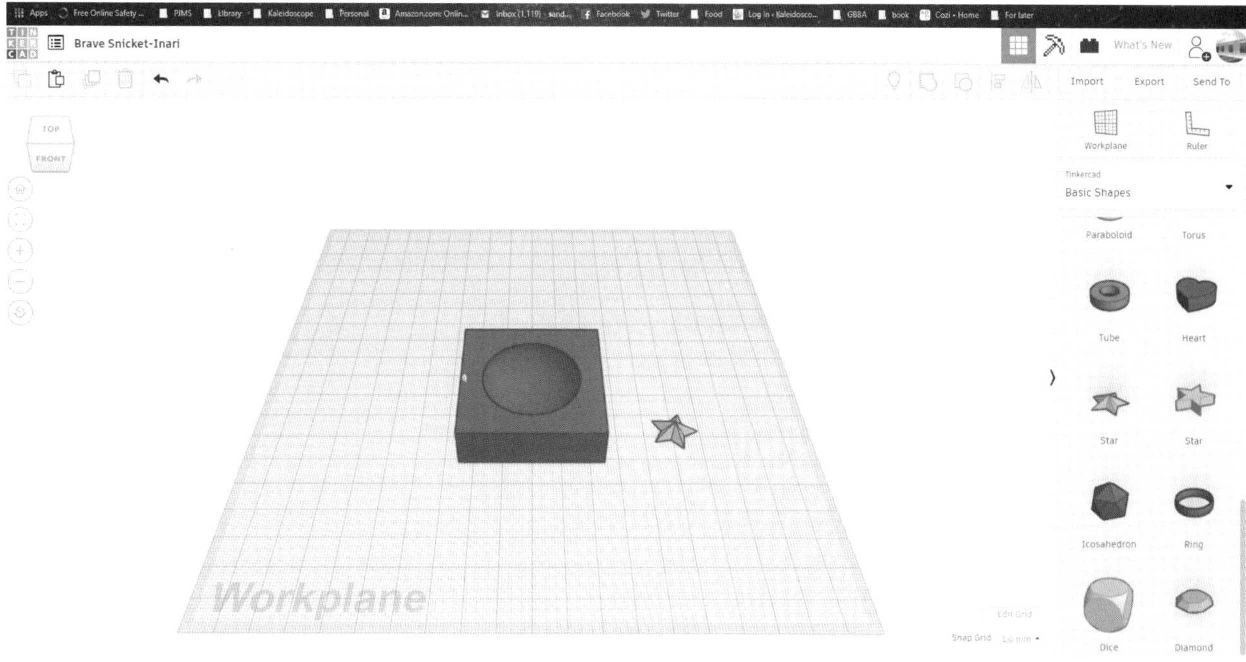

FIGURE 6.14

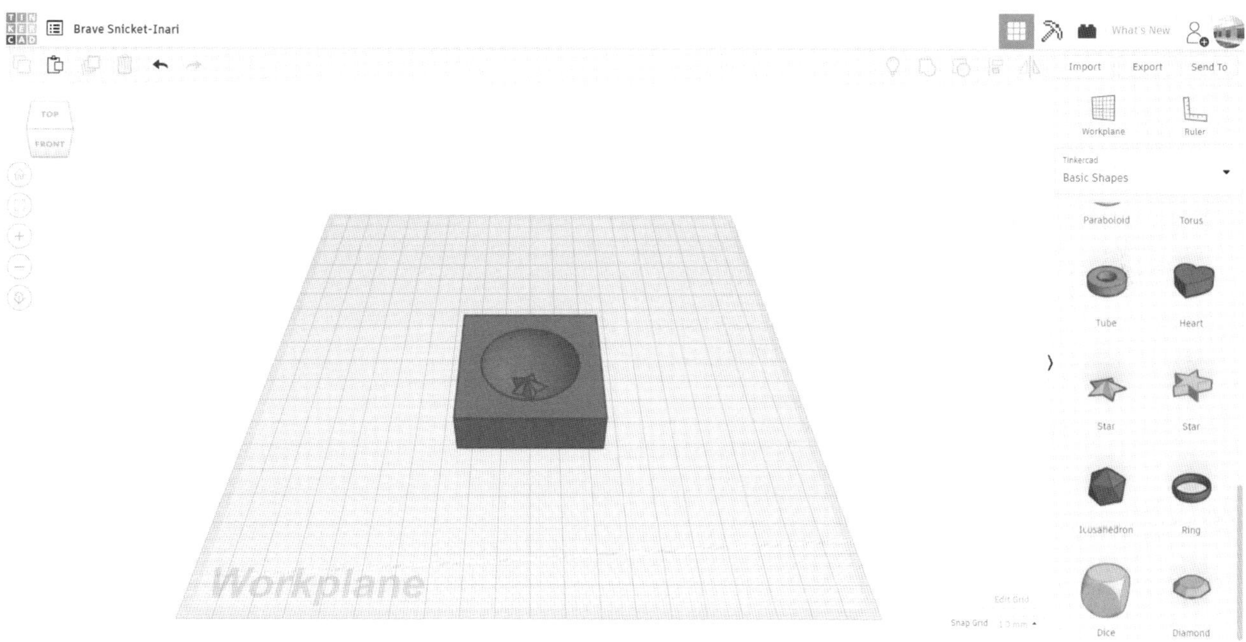

FIGURE 6.15

14. Gently tap the mold to remove any air bubbles.
15. Chill the chocolates until solid (10–20 minutes in the refrigerator).
16. Unmold the chocolates by inverting them and gently smacking the mold against a cutting board or the cookie sheet. If the chocolates don't release, let them warm up a bit, or place the bottom of the mold in warm water. Then try inverting them again (Figure 6.16).

FIGURE 6.16

TIP: Want to mold a more complicated item? Search Thingiverse.com for free 3D-printed objects that you can use. Import an object and set it as a "hole." Use the "Flip" tool as needed to orient the object, and then set it into a "box" to create a mold.

PROJECT 3
3D-Printed Cookie Cutters

COST: $$$

MAKE TIME: 15–30 minutes to design, time varies to print

Baking is such a fun thing to do. Even simple sugar cookies, cut and decorated, allow for tremendous creativity and celebration. So why not make camp cookies? To make it even more fun, let your campers make their own cookie cutters, adding another level to the design work.

For this activity, campers will need crisp black-and-white designs. The designs should not have too much detail because that would make it difficult to cut cookies. The image should have a solid white background without a pattern. Looking for images gives camp leaders a great opportunity to show campers how to use the advanced image search tools on Google, which allow users to select the colors and sizes of the images. Campers can also use software such as Corel Draw, Photoshop, Illustrator, or GIMP to create images.

If you want a low-tech design method, simply give campers copy paper and black markers to draw their shapes, then scan or photograph the images, and import them to the computer. For a personal approach, use a bright light and white paper (like the kind used for bulletin boards) to trace silhouettes of the campers then transform them into digital images. There are also apps available to help you take silhouette images (Figure 6.17).

For more information on 3D printing and choosing software, see Chapter 3. You will need to select CAD software such as Tinkercad, SketchUp, or Morphi for this project. Because it is free and easy to use, the projects in this book are demonstrated using Tinkercad.

To import a graphic into Tinkercad, it must be an SVG file. SVG stands for Scalable Vector Graphics and is a vector image format for two-dimensional images. You can use an online converter such as Online-Convert (https://image

FIGURE 6.17

.online-convert.com/convert-to-svg) to do the work. Of course, if you are using computer software to create or modify your images, you can often save them into this format directly. Vector arts editors such as Adobe Illustrator allow you to save in this format, as does the free, open-source Inkscape. Apps such as Morphi have their own image-conversion features built in.

If you don't want to design your own cookie cutter in CAD software, you can use Cookie Caster online (http://www.cookiecaster.com/). This free website lets you import images or digitally draw your own. It renders the STL print file, which you can download. Especially with younger campers or larger groups, this is a great option that allows everyone to generate a file quickly and easily. Those who want to learn more about graphic design and CAD software, however, will appreciate working from scratch (Figure 6.18).

As for filament, you absolutely need to use PLA. This material is considered to be nontoxic. For best results and best safety, select a food-grade or food-safe filament. Many options are available. If you are unsure, contact the seller. If you are unable to get food-safe filament, you can use standard PLA and cover it with kitchen plastic wrap. For a more detailed discussion on the food safety of 3D-printed objects, see Section 2 on chocolate making.

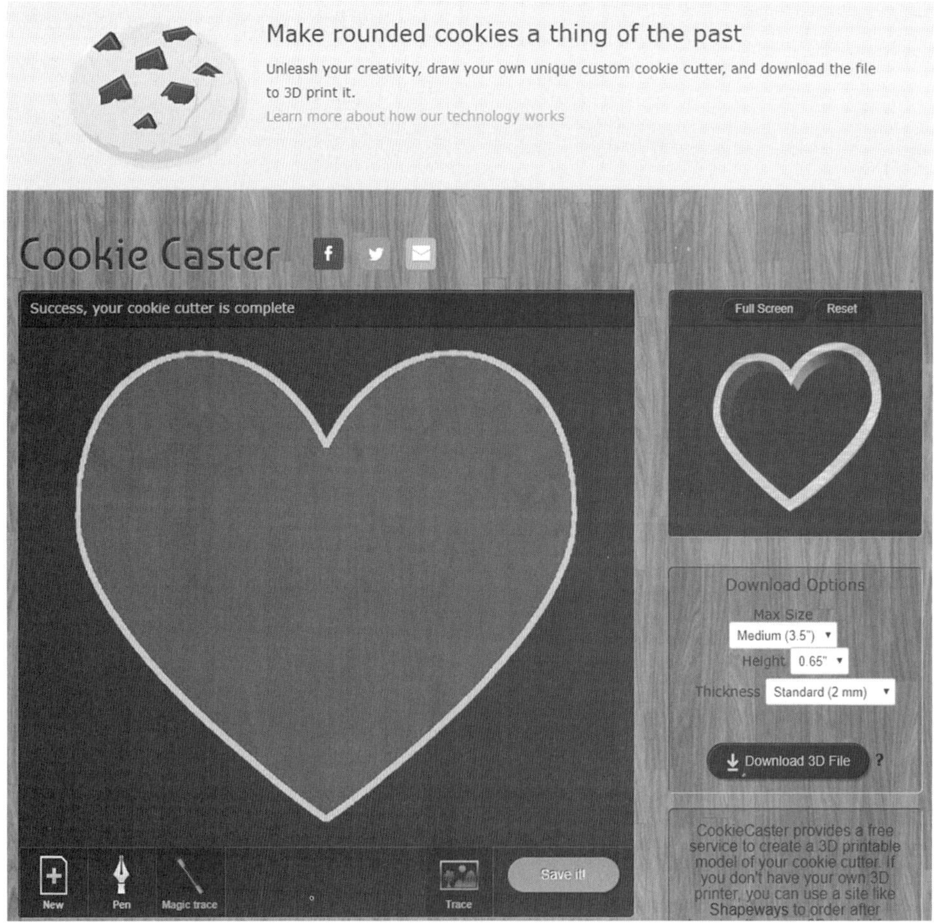

FIGURE 6.18

Materials

- A black-and-white image
- CAD software, such as Tinkercad
- Graphics arts software, such as Photoshop or GIMP
- Vector arts editor, such as Illustrator or Inkscape
- PLA filament, food safe
- 3D printer of your choice
- Cookie dough (see recipes below)

Process: Designing a Basic Cookie Cutter in Tinkercad

Simple symmetrical shapes work best for this method. Any shape with too many twists and turns will be difficult.

1. Either use the Scribble feature to create a drawing or use shapes to create a design. Alternatively, import an SVG file using the "Import" feature. Once grouped, click on the white box in the corner of the design, and drag it to size the object to roughly 89 millimeters (3.5 inches) in height, which is a standard medium cookie cutter. Make note of the width. Click the white box on top of the design, and drag it to size the object to a depth of 13 millimeters (½ inch).
2. Copy the object using either "Copy" (CTRL+C) and "Paste" (CTRL+P) or "Duplicate" (CTRL+D). Move the second object to the side. Repeat this action two more times so that you have four shapes (Figure 6.19).
3. Select one of the shapes, and resize it to be 2–3 millimeters smaller than the original design using the white boxes. It may be easier to click on the size number shown and directly enter the number rather than dragging the image. Check that the depth of the object is no more than 13 millimeters. In the shape control box, click "Hole." The object should turn gray and white striped (Figure 6.20).

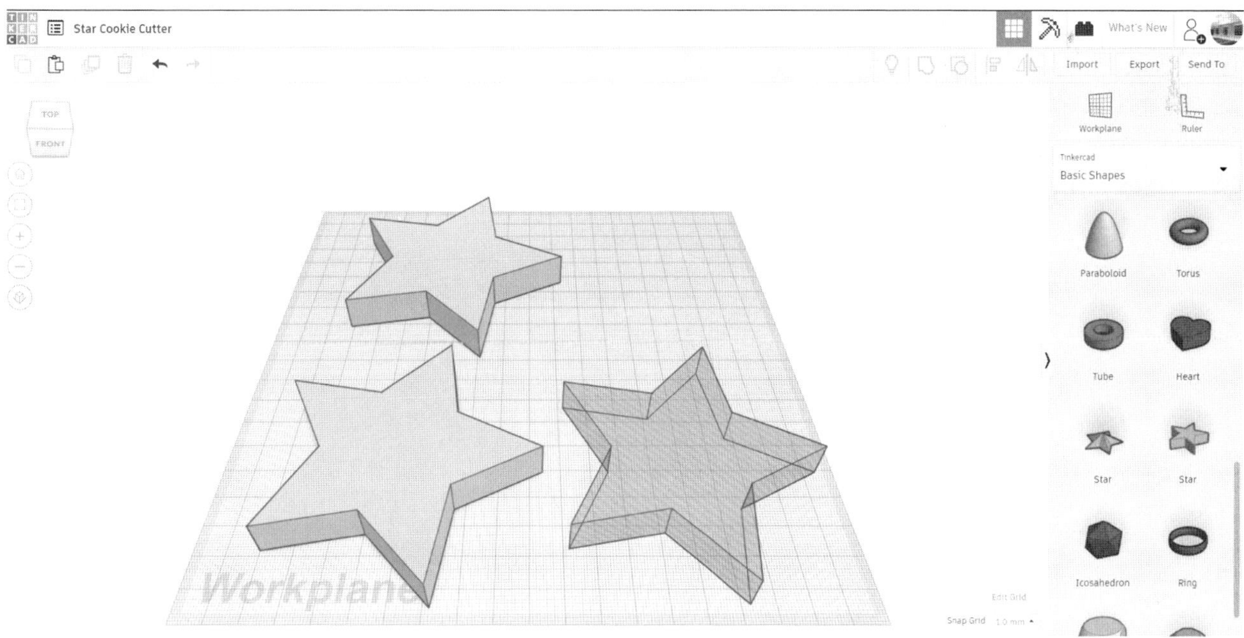

FIGURE 6.19

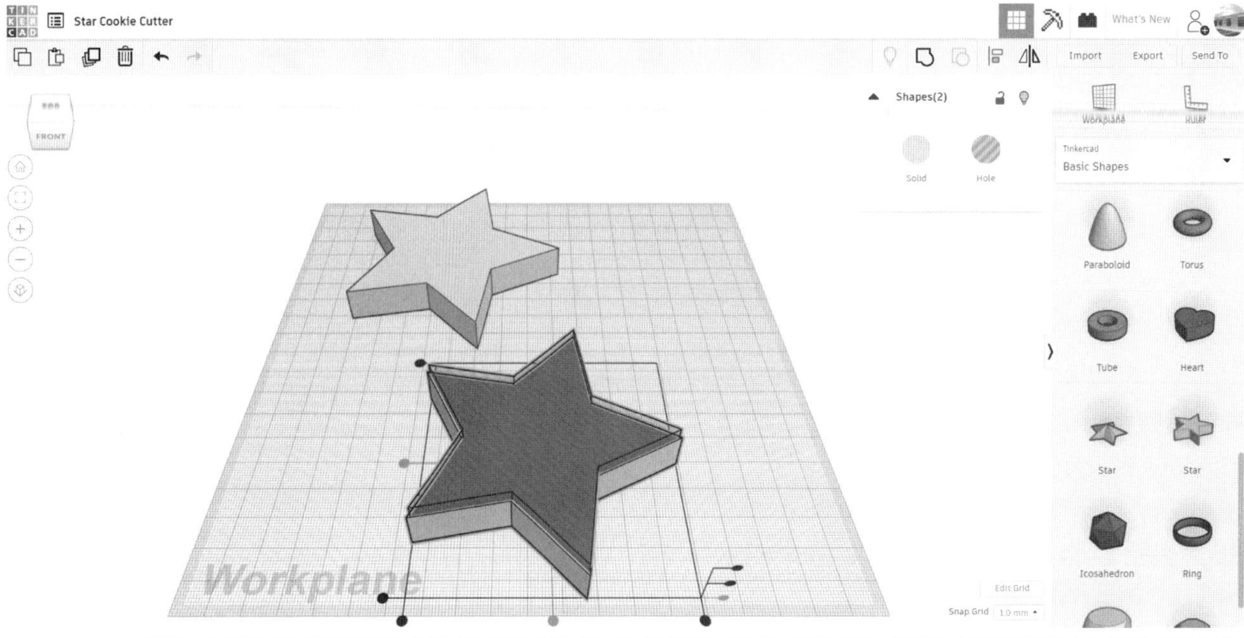

FIGURE 6.20

4. Move the hole object over the original object. You may need to adjust the view to make sure that the two objects are lined up properly. The goal is to center the hole object in the center of the original. It can be helpful to select both objects and use the "Align" feature in the upper right of the screen to line up the objects. When they are arranged properly, group the objects. You should now have a thin outline of the shape that will serve as the cutter (Figure 6.21).
5. Select one of the copies of the original, and resize it to be 2–3 millimeters larger than the original, as described in step 3. Select the

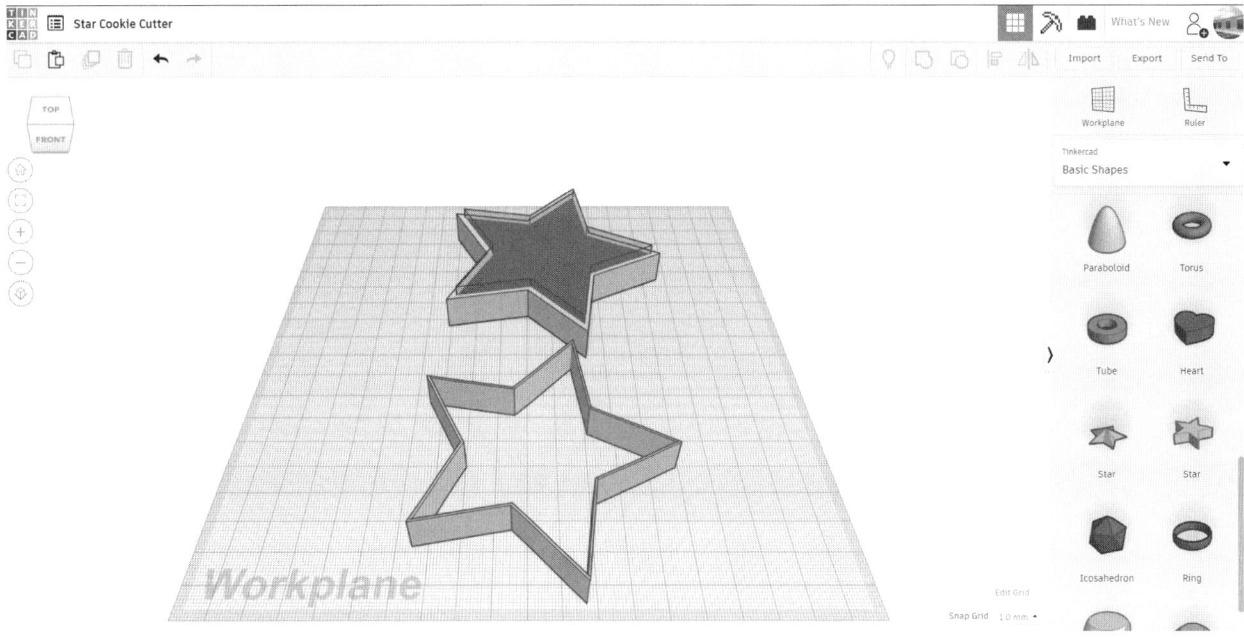

FIGURE 6.21

last copy, and size it to be 5–6 millimeters smaller than the original. Make this object a “Hole.” Using the same method as in step 4, move the hole over the enlarged object and group them. Using the white box on top, resize the depth of the object to 2 millimeters. This is the rim or base of the cookie cutter, which will make it easier to grip (Figure 6.22).

6. Move the rim to the cutter, and line it up so that the cutter is roughly centered on the rim. If needed, use the “Align” tools. Group the objects (Figure 6.23).
7. 3D-print the object—and make cookies!

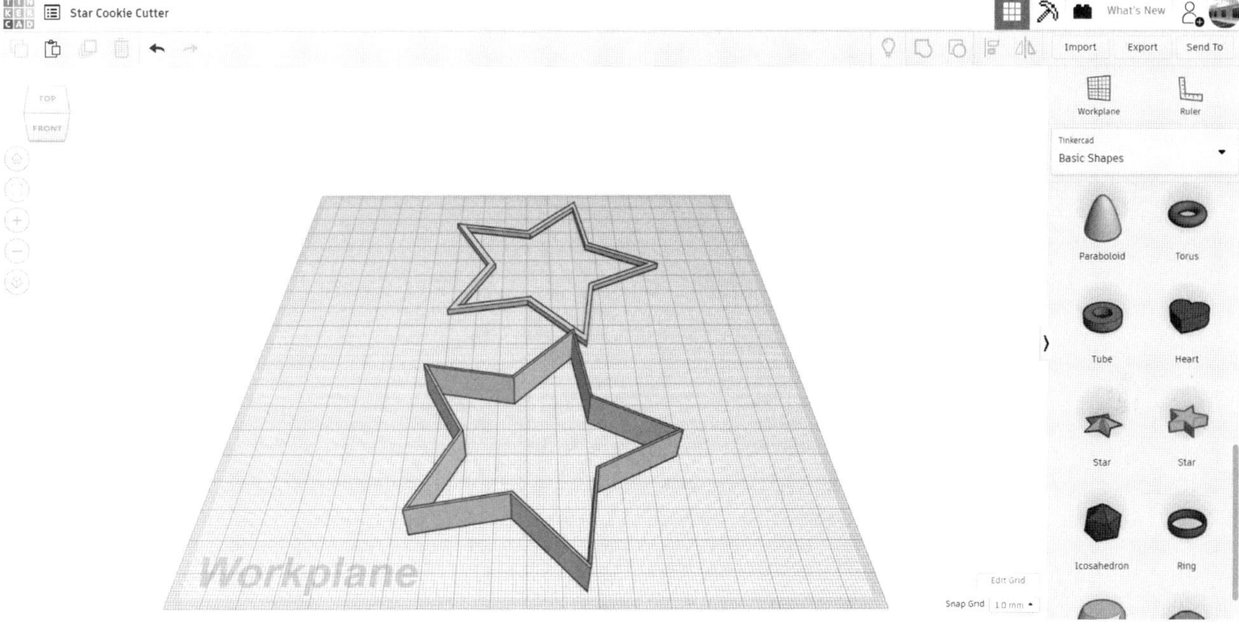

FIGURE 6.22

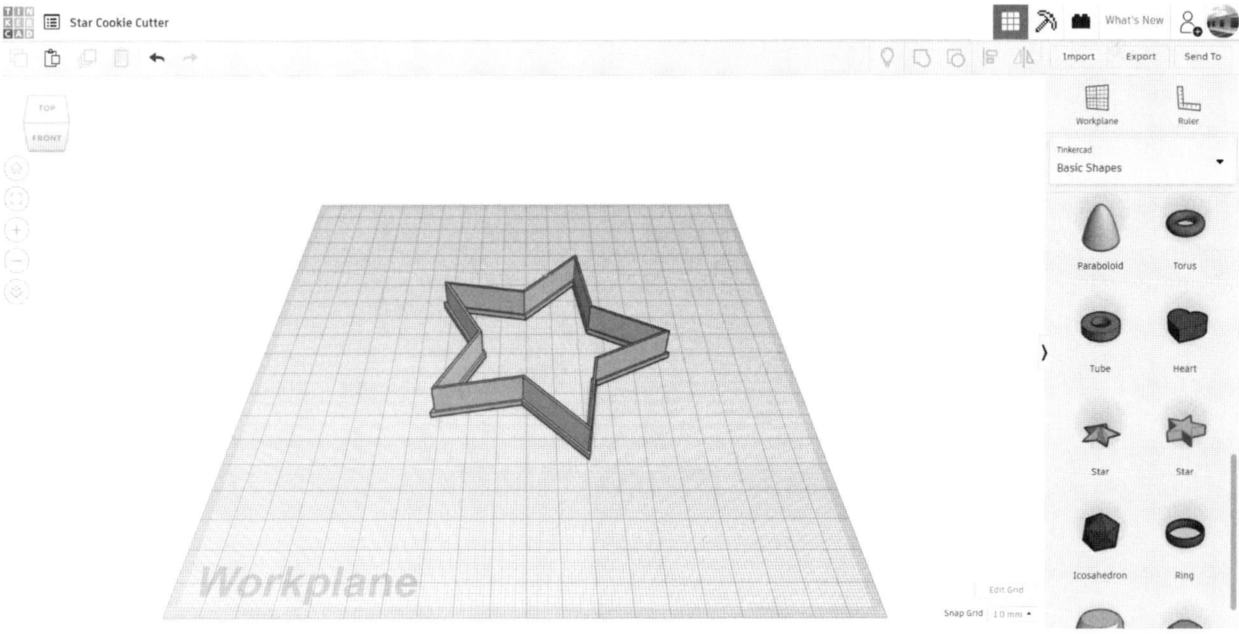

FIGURE 6.23

Process: Designing a Basic Cookie Cutter in Illustrator/Inkscape

This method uses a hybrid of vector arts editing software and CAD software to create cookie cutter pieces. It will let you use more complicated shapes and create cookie cutters with an imprint or embossed design. You can save directly from vector arts editing software into SVG files, ideal for importing in to Tinkercad. Because this software is free and open source, I am using Inkscape for this project, although the process is very similar with Illustrator. Download Inkscape at https://inkscape.org/.

1. Open a JPG or PNG file in Inkscape.
2. Select the image. Go to the "Path" menu, and choose "Trace Bitmap." Set colors to "2," and click the "Remove Background" box (Figure 6.24). Separate the new vector trace from the original image, and then remove the original (Figure 6.25).
3. From the "Path" menu, choose "Simplify" (Figure 6.26).

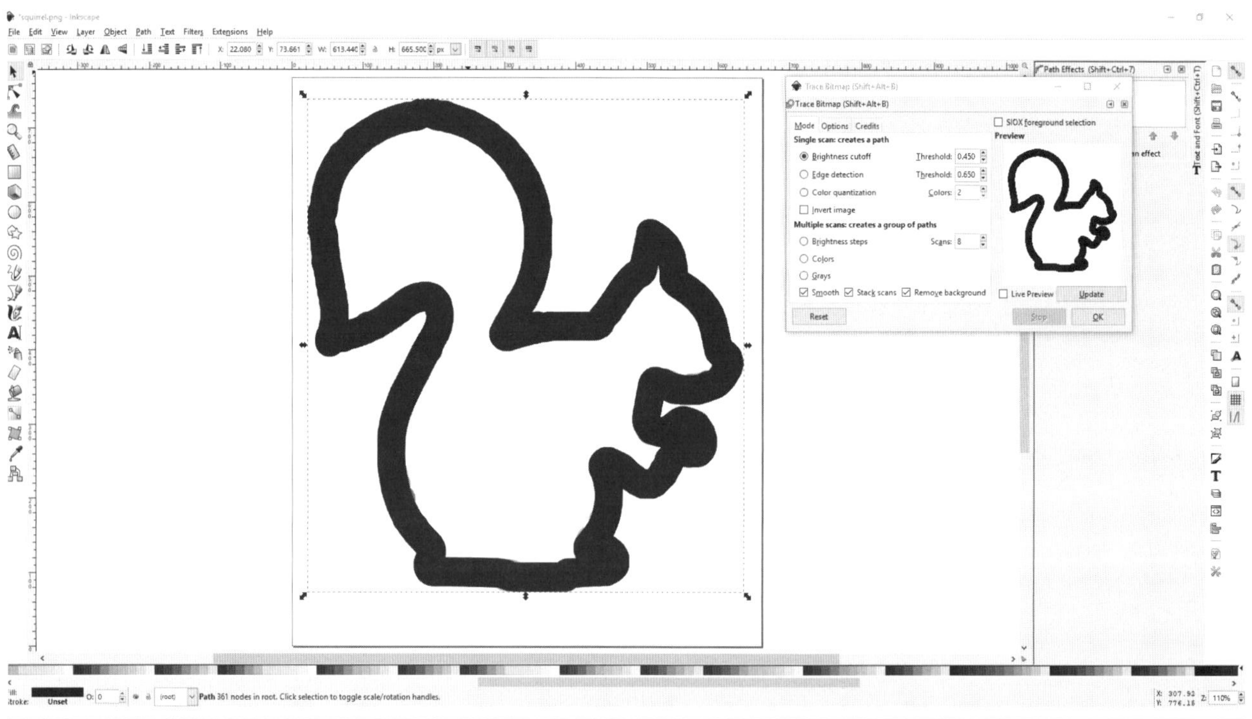

FIGURE 6.24

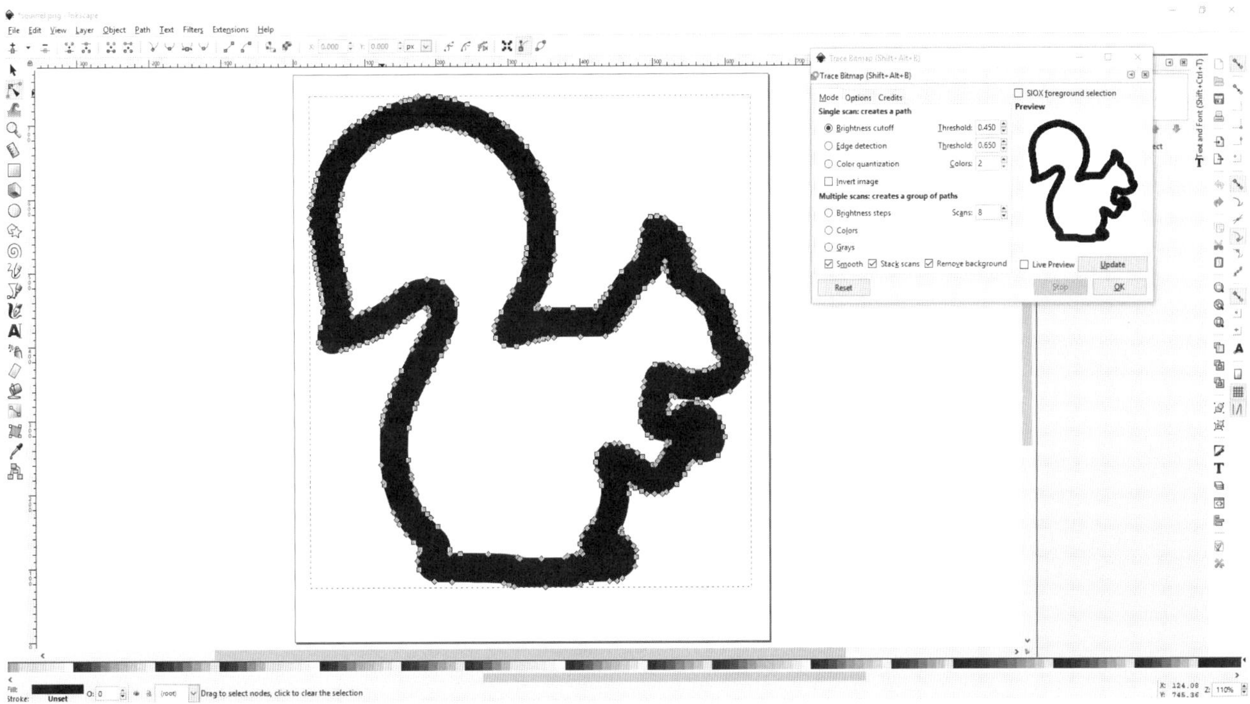

FIGURE 6.25

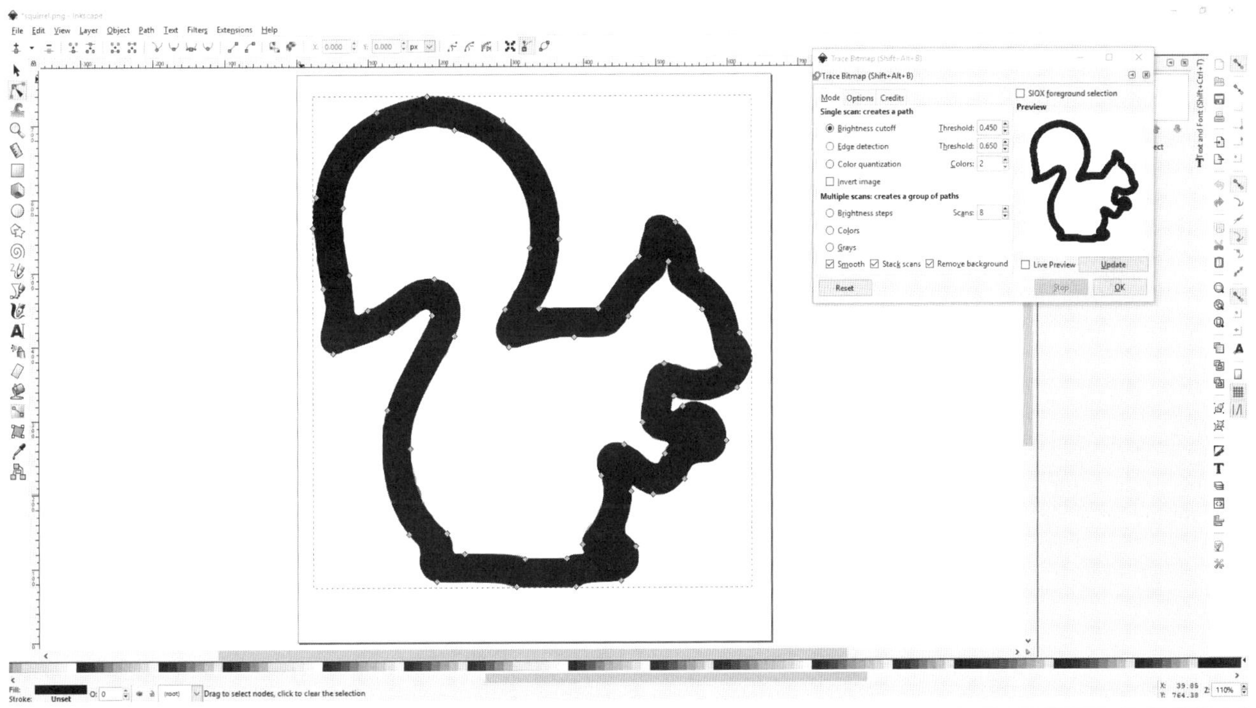

FIGURE 6.26

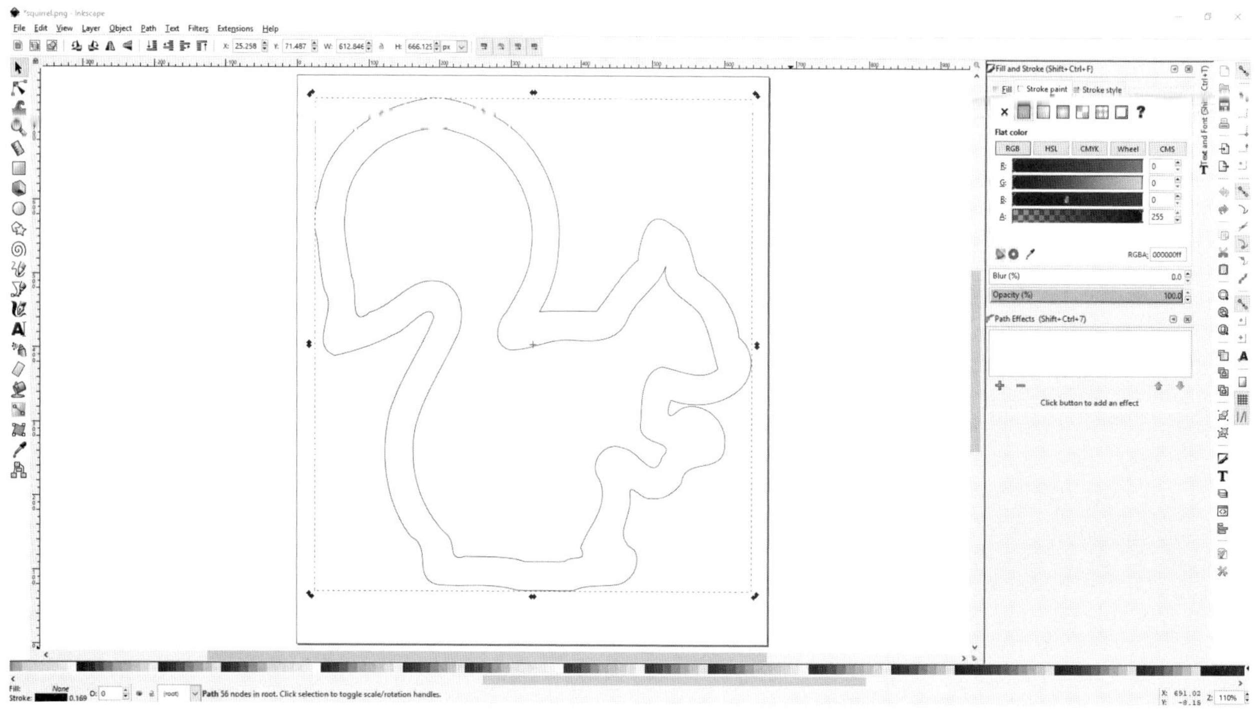

FIGURE 6.27

4. From the "Object" menu, select "Fill and Stroke." Set the fill to invisible (the "X") and the stroke to black. You should now have a clean outline of your object (Figure 6.27). Note that if you have an outlined drawing (like I do in this example), you may need to use "Path" and then "Break Apart" to separate the two outlines. Select one outline, and remove it using "Delete" (Figure 6.28).

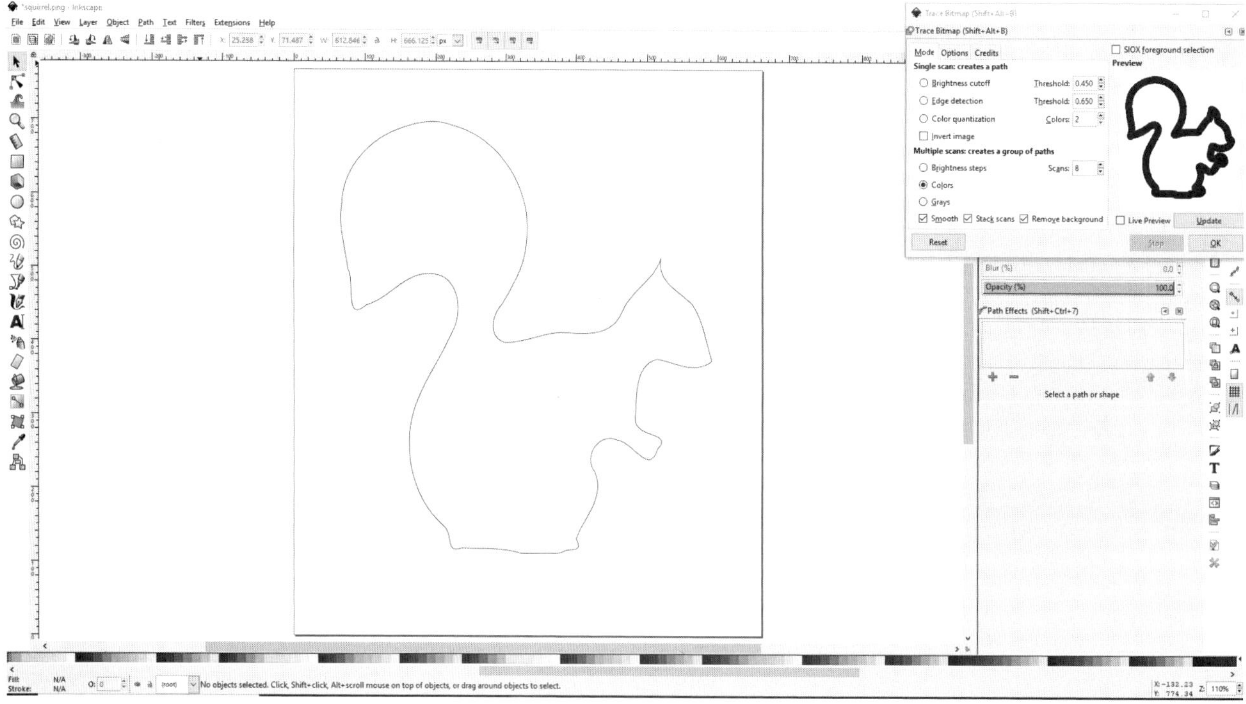

FIGURE 6.28

5. Select the object, and copy it. From the "Path" menu, select "Outset" (CTRL+)). Do this several times (8–10 times should give you a good base).

NOTE: In Illustrator, you will select the "Object" menu, then "Path," and then "Offset Path." Set the offset to 0.3 inch (Figure 6.29).

6. If needed, from the "Object" menu, select the "Align and Distribute" tool. Select both objects. Then use the "Center in Vertical Axis" and "Center on Horizontal Axis" tools to align the items. Save the file as a "Plain SVG" file *not* an "Inkscape SVG" file. This will be the base of the cookie cutter.

7. Delete the larger object, and repeat step 5, but this time only use "Outset" two to three times. Repeat step 6, naming the file differently. This will be the cutter portion of the cookie cutter.

8. Go to Tinkercad, and use "Import" to bring in both files. You may need to resize them to import them. If you do need to resize, change the scale rather than the height or width. I usually set the scale to "10."

9. Using the white box on top of the base object, set the depth to 2 millimeters. This will be the rim that makes the cutter easy to hold. Set the depth of the cutter object to 13 millimeters (Figure 6.30).

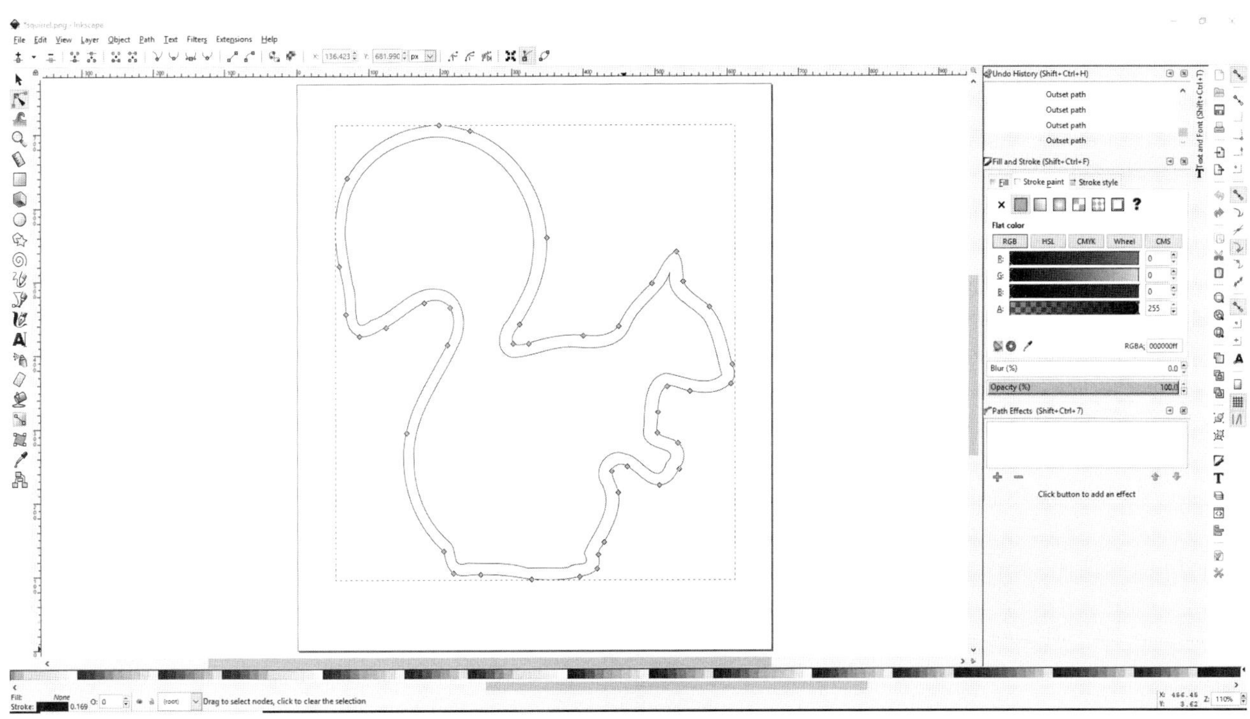

FIGURE 6.29

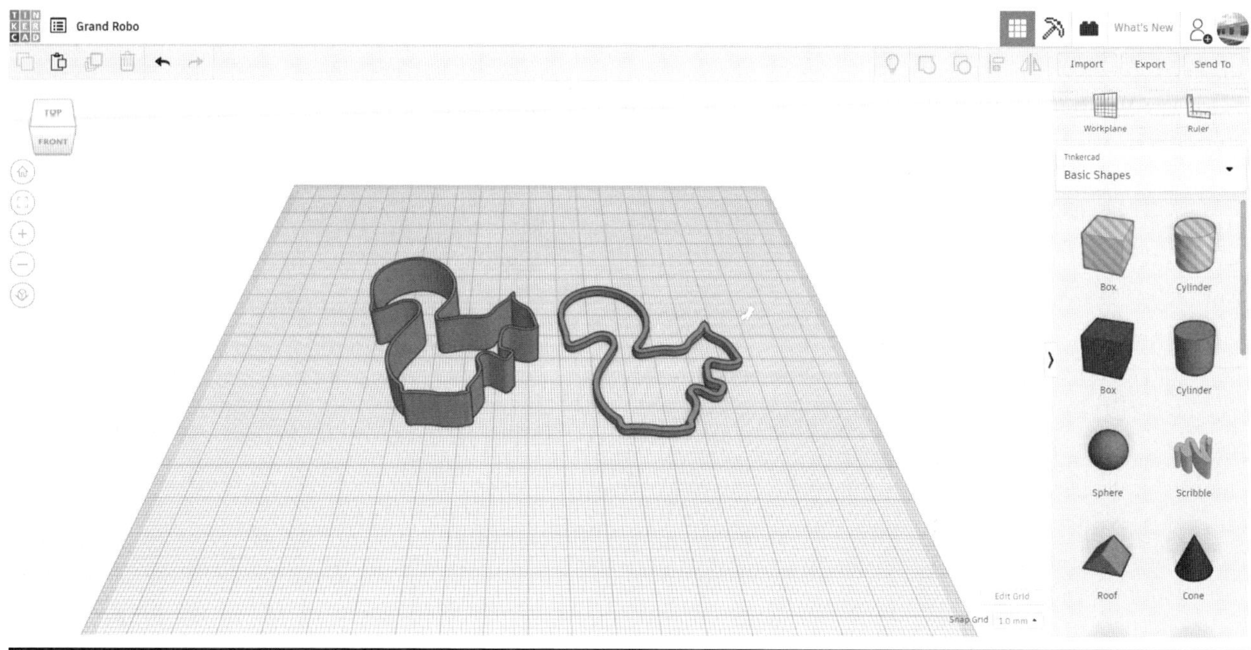

FIGURE 6.30

10. Move the rim to the cutter to line up with the cutter so that the cutter is roughly centered on the rim. If needed, use the "Align" tools. Group the objects (Figure 6.31).
11. 3D-print the object—and make cookies!

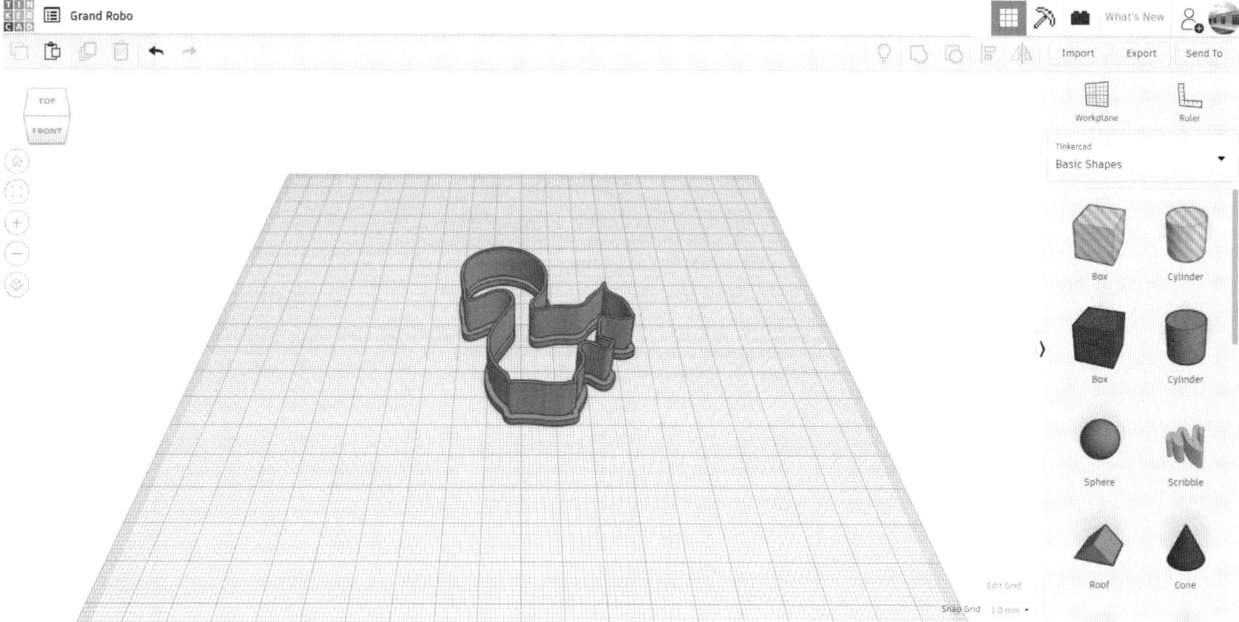

FIGURE 6.31

Process: Designing a Basic Cookie Cutter in Photoshop/GIMP

This is my favorite method for creating cookie cutters because using the "Border" tool allows you to make a base that aligns better with the cutter portion of the cookie cutter, especially for complex designs. It also allows you to use line art drawings to make imprinted and embossed cookies easily. However, you cannot save the file as an SVG file directly from a graphics arts program, so you will need to convert your file elsewhere. Because it's free and open source, I'm using GIMP for this project. The process is similar in Photoshop or CorelDRAW. Download a copy at https://www.gimp.org/.

1. Open your black-and-white image in GIMP. Use the "Magic Wand" tool to remove any background. Use the "Resize Image" tool under the "Image" menu to resize the image to 3½ inches high. Make sure that the chain icon is unbroken so that the height and width change together. If there is no space around the image, use the "Canvas Size" tool from the "Image" menu to create space around your image. Use the "Magic Wand" tool to select the black part of your image (Figure 6.32).

FIGURE 6.32

2. From the "Select" menu, choose "Border." Set the border to 0.1–0.3 inches depending on how large you want the base or rim of the cookie cutter. This may also vary based on the size and resolution of your image. Make sure that the border is set to "Smooth." Hit "OK" (Figure 6.33).

3. From the "Layer" menu, select "New Layer." Create a new layer using the default settings. From the "Edit" menu, select "Fill with FG Color." Click on the eye icon for the original layer so that it is hidden. You should just see the outline now (Figure 6.34).

FIGURE 6.33

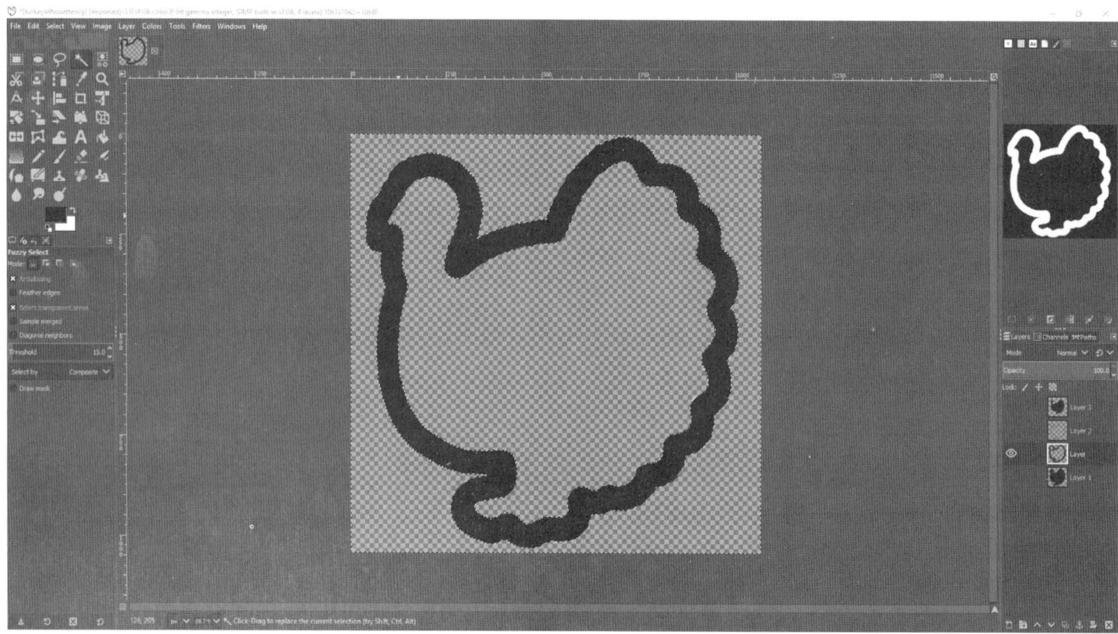

FIGURE 6.34

4. Under the "File" menu, select "Export," and save the file as a JPG or PNG file.

NOTE: Although Adobe has a save to .SVG function, I did not have good success saving files this way for use in Tinkercad (Figure 6.35).

5. Hide the outline, and click on the eye icon to show the original image. Repeat steps 2 and 3, but set the border to be much thinner, roughly half of what you did previously. Save this file under a different name.
6. Convert your files to SVG files. Either use the online tool mentioned earlier or use Inkscape. In Inkscape, open the file. Select the image, and then from the "Path" menu, select "Trace Bitmap." Check the box next to "Edge Detection" and "Remove Background." Click "OK." Remove the original image, and keep the traced path. From the "Path" menu, select "Simplify." Save as a plain SVG file. Do this for each file.
7. Go to Tinkercad, and use "Import" to bring in both files. You may need to resize them to import. If you do need to resize, change the scale rather than the height or width. I usually set the scale to "10."
8. Using the white box on top of the base object, set the depth to 2 millimeters. This will be the rim that makes the cookie cutter easy to hold. Set the depth of the cutter object to 13 millimeters.

FIGURE 6.35

9. Move the rim to the cutter and line it up so that the cutter is roughly centered on the rim. If needed, use the "Align" tools. Group the objects (Figure 6.36).

10. 3D-print the object—and make cookies (Figure 6.37)!

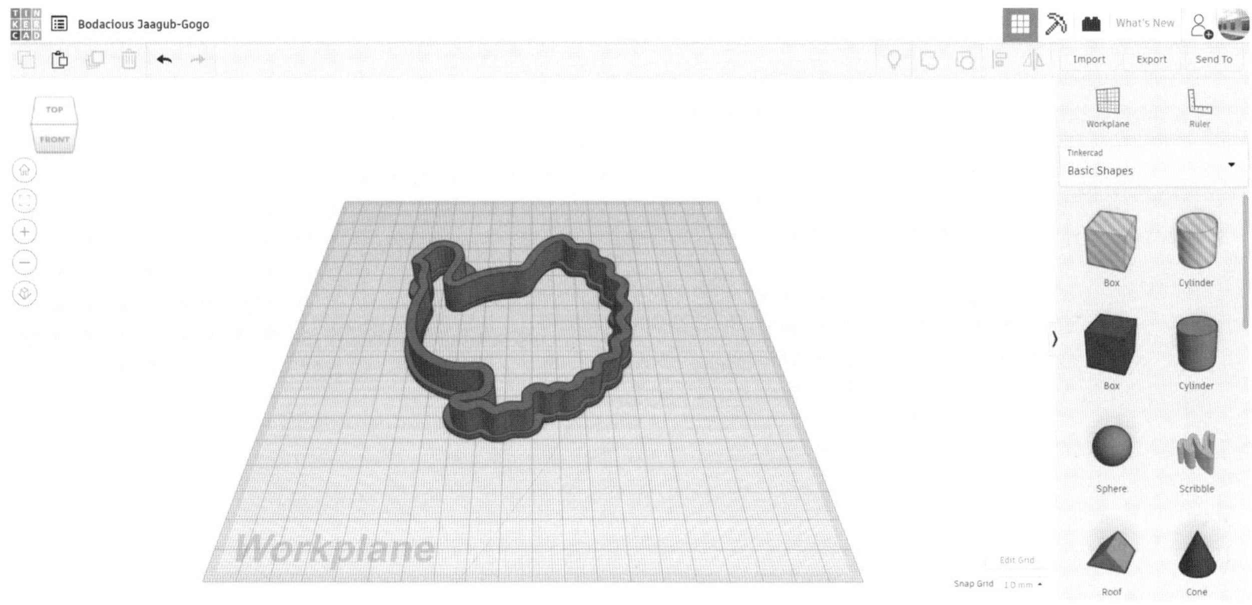

FIGURE 6.36

FIGURE 6.37

Take It Further: Making Imprinted or Embossed Cookie Cutters

The process for creating embossed cookie cutters is very similar to that for creating the basic cutouts. The internal embossing piece should have a depth of at least 7 millimeters for a 13-millimeter cutter. This will allow the internal sections to be pressed onto the face of the cookie dough. Simple coloring pages can provide great line art for embossed cookie cutters.

Rather than an open base, you can simply make a solid piece that is 2–3 millimeters larger than the cookie cutter object. Attach the internal items to that base, and you'll have a sturdy embossing cutter. If you are concerned about printing time or the amount of filament used, you can create bands that go across the object and attach to the base/rim to hold the embossable portions of the design in place.

If you are working in Tinkercad, you can simply create the internal patterns or images you want for embossing and set them to the appropriate depth of 7 millimeters. When working in Inkscape or GIMP, you will create the embossed sections as a third SVG file. The trick is to select only the parts of the original image you want for each portion. Play with the settings in your software to do this. For example, in GIMP, you can use the "Magic Wand" tool to select outside the image and then the "Invert" tool from the "Select" menu to create the border for the cutter (Figure 6.38).

FIGURE 6.38

Process: Making the Cookies

Everyone has their favorite recipe, but this is my favorite sugar cookie recipe. They cut well and keep their shape when baked. I've substituted the all-purpose flour with commercial gluten-free blends successfully as well, so it's very versatile. Decorate the cookies with sprinkles and colored sugar before baking or with royal icing afterwards, as desired.

Ingredients

- 1 cup unsalted butter, softened
- $\frac{2}{3}$ cup sugar
- 1 large egg
- $\frac{1}{4}$ teaspoon baking powder
- $\frac{1}{8}$ teaspoon salt
- 1 teaspoon vanilla
- $2\frac{1}{3}$ cups all-purpose flour

Instructions

1. In the bowl of a stand mixed, combine sugar and butter and mix until light and fluffy. Add the egg and vanilla and blend well.
2. In a separate bowl, combine the dry ingredients. With the mixer set on low, gradually add the flour mixture until it is just incorporated. Scrape down the bowl and mix one last time.

3. Make a ball of the dough, and place the dough on plastic wrap. Flatten the ball and wrap well. Place the dough in the refrigerator for 30–60 minutes until chilled well.
4. Preheat the oven to 350 degrees F.
5. Place each circle of dough between large sheets of wax paper. Roll the dough to a scant ¼ inch thick, checking the underside frequently and smoothing out any creases. Use the cookie cutters to cut shapes as desired.
6. Place cookies on a baking sheets lined with parchment paper. Bake for 6–9 minutes, or until the cookies are just slightly colored on top and darker at the edges. Rotate the baking sheet halfway through the baking for even browning.

NOTE: Gluten-free cookies will not brown as much as traditional cookies.

7. Once the cookies are removed from the oven, let them firm up (1–2 minutes). Then transfer the cookies to wire racks and let stand until thoroughly cool.
8. Makes about 3 dozen 3½-inch cookies.

PROJECT 4
Solar Ovens

COST: Free–$

MAKE TIME: 15–30 minutes

Cooking with the power of the sun is not a new idea for Maker Camp, but some things are classics for a reason. Cobbling together a cardboard box with foil and plastic wrap to create a solar oven is a solid engineering and maker activity. The science behind why it works is important, and the delicious treats you get to enjoy as the fruit of your labor are pretty awesome.

Let's talk about construction first. You'll need a box made of corrugated cardboard. Thinner materials won't hold the heat well, so skip the cereal boxes. You don't want too large a box because that just means you have to heat unnecessary air to cook your food. I prefer flip-top boxes, which make it easy to add and remove items. Pizza boxes and postal mailers are great choices and can be found or ordered easily. You may find that your favorite local pizza place or nearby post office is happy to donate boxes. If you don't have the same box for every camper, consider it a great opportunity to engineer, prototype, and compare results.

Solar ovens are a great opportunity to demonstrate sunlight-to-heat conversion. As the electromagnetic waves of light hit matter, that energy is transferred to the atoms and molecules in the oven itself, the food, and the air inside. Those particles start to vibrate more quickly as a result, and this creates what we know as heat.

This happens anywhere sunlight touches, but a solar oven is designed to make the best use of the heat that's generated. The box should be sealed as tightly as possible to keep the heat inside. The black paper on the inside of the box helps to absorb as much sunlight as possible, generating more heat. A plastic window allows the light in while trapping heat, much as our own atmosphere does, and the foil-covered panel helps to reflect sunlight into the box for the best effect.

When it comes to cooking, there are a few tried and true items to prepare. S'mores are usually a camp favorite. You won't get the same toasty goodness as you would from a campfire, but the gooey melted sweetness is still satisfying. Pizza toppings on an English muffin and a small tray of tortilla chips covered in cheddar cheese are great savory items that are easy to make.

Materials

- Cardboard boxes
- Black construction paper
- Plastic wrap
- Aluminum foil
- Bamboo skewers, pencils or wooden dowels
- Masking or invisible tape
- Glue sticks (optional)
- Scissors
- Box cutters, craft knives, or cardboard cutters
- Safety equipment such as gloves and goggles
- Small plates
- Food to cook

Process

1. Remove any debris from your box as needed. Using you cutter, cut an opening in the top of the box that takes up most of the space, leaving enough of an edge to retain strength. Only cut three sides, so that you can flip this panel of cardboard up and away from the top of the box.
2. Using glue and tape, cover the bottom side of the flip-up panel with aluminum foil. Do your best to keep the foil as flat and smooth as possible.
3. Using glue or tape, cover the inside bottom and sides of the box with black construction paper.
4. Cover the hole you made with plastic wrap. Try to keep the plastic taut and smooth. Use tape to completely seal around the edges of the plastic wrap.
5. Prepare your food on a small paper plate. Lift the top or side of your box carefully. Slide the plate inside the solar oven.
6. Bring the oven outside, and find a sunny location. Face the foil-covered panel towards the sun and angle it until as much light as possible is reflected into the plastic-covered hole. Use the bamboo skewer and some additional tape to secure the panel at this angle.
7. Allow the oven to sit for up to a half hour in the sun, checking it regularly until the marshmallow, chocolate, or cheese has melted (Figure 6.39).

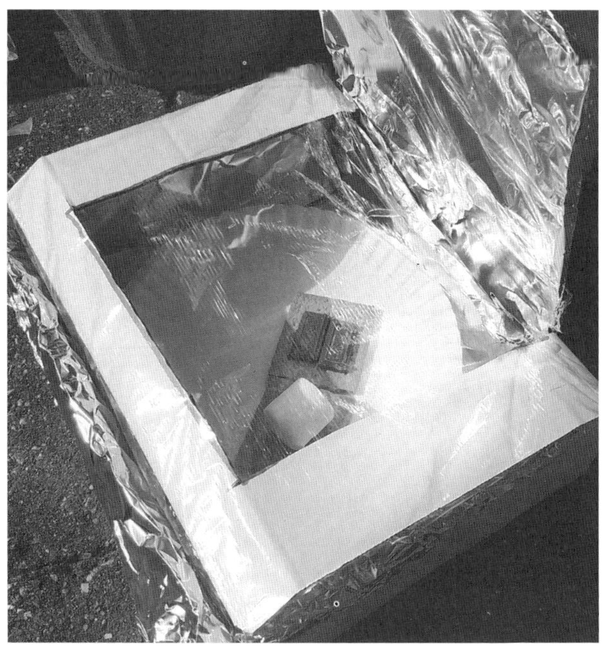

FIGURE 6.39

Take It Further

Sometimes campers don't realize how warm it can get inside the oven. Try placing a thermometer outside of an oven and another one inside so that campers can see the how much heat the sun's light can create.

Campers may want to decorate the outside of the oven. Ask them what colors would be best for generating heat. They can also try adding insulation outside the box or lining the inside of the box with different materials to see if they affect the results.

While things are cooking, have campers place thermometers under differently colored construction paper to determine how the colors affect how much heat is generated. Or try placing thermometers in glass mason jars. Add an inch of soil to one, an inch of sand to another, an inch of gravel to another, an inch of water to another, and nothing to the last. Seal the jars, and place them in the sun. Do the materials in the jar affect how much heat is generated?

PROJECT 5
Edible Treehouse

COST: $$

MAKE TIME: 15–30 minutes

This is a fun edible engineering challenge based loosely on the idea of building gingerbread houses. Because it is a very flexible activity, you can adapt it to anything you want. For example, one year when I held a superhero-themed Maker Camp, we all created fortresses for our heroes or lairs for our villains. Alternatively, you could make this activity a simple challenge, such as building the tallest tree house you can with the materials provided.

You can do this activity with just graham crackers, icing, and candy, but adding other types of cookies, crackers, and other materials makes it more interesting. It gives campers a chance to test the qualities of each material for strength, brittleness, and flexibility. It also gives campers the opportunity to build more creatively. So have fun with your materials selection!

As for icing, traditionally, this activity uses royal icing, which hardens. The downside is that you must make this icing yourself from scratch on the day you plan to use it, usually within hours of use. This can be a fun activity to do with your campers, so I provide the recipe after illustrating the activity.

However, you also can use store-bought icing for this challenge. It won't set up as a hard cement-like material, but it will allow campers to stick pieces together. You can provide toothpicks or gummy candies to add additional support. If you use store-bought icing, splurge on the kind that comes in tubes. It'll make your life, and cleanup, a lot easier. Otherwise, use piping bags from the craft store or zip-top bags with a bottom corner snipped off to apply the icing.

Materials

- Graham crackers
- Assorted cookies, such as sugar wafers, sandwich cookies, etc.
- Assorted snack crackers
- Assorted snacks such as pretzel logs, cheese balls, etc.
- Assorted candies, both hard and gummy
- Icing of your choice
- Toothpicks and/or bamboo skewers (optional)
- Popsicle sticks
- Paper plates or cake boards

Process

1. Give each camper a paper plate or cake board on which to build. This gives everyone a stable base to start with.
2. Have campers select their materials.
3. Pass out icing, and begin building. Use toothpicks or bamboo skewers to reinforce as needed. Use Popsicle sticks to smooth the icing as needed (Figures 6.40 and 6.41).

FIGURE 6.40

FIGURE 6.41

Royal Icing Recipe

Ingredients

- 3 tablespoons meringue powder
- 4 cups confectioners' sugar
- 6 tablespoons warm water

Instructions

1. Beat all ingredients until icing forms peaks (7–10 minutes at low speed with a heavy-duty mixer, 10–12 minutes at high speed with a handheld mixer).
2. Thin and color the icing as needed.

PROJECT 6
Campfire Popcorn

COST: $

MAKE TIME: 5–15 minutes

When I was a kid, we went camping in the Catskill Mountains of New York every summer. My family has gone there for over 80 years now. Camping is in our blood. My sister and I continue the tradition, taking our own families to upstate New York every year to enjoy this family tradition. We all look forward to it.

Of course, the family joke is that if we are camping, it must be raining. For decades now, we have not managed to avoid multiple days of rain on our camping trips. Preparing for the rainy days has become a point of family pride.

This is why I'm in love with Jiffy Pop. When the weather got bad, my family and I would hunker down in our screen house, set over a camp picnic table, and play cards. As it got dark, Dad would make Jiffy Pop and Swiss Miss hot cocoa, the kind with mini marshmallows. Listening to the corn kernels pop as the foil top expanded was so exciting. This was literally a once a year occurrence, so my sister and I treasured it.

These days, it's not so easy to find Jiffy Pop popcorn. Unless you live in a "camping" area, stores simply don't carry it. If they do, it isn't always inexpensive. And even if it's inexpensive, it's only available during the summer. Nowadays, microwave popcorn is all the rage.

But I still wanted to share that popcorn experience with my kids when we were camping and all year long. Luckily, it's not hard to re-create the Jiffy Pop experience. With a little planning, you can make popcorn over a fire, a camp stove, or a regular kitchen stove. It's physics in action. What's not to love?

When determining how much popcorn to make, keep in mind that 2 cups of popped popcorn is considered a serving. So ¼ cup of dry kernels will yield four servings, if most or all of the kernels pop. With the method outlined below, you should have minimal burned and few unpopped kernels.

Feel free to get creative with the seasoning, but don't add it until after the kernels have popped. You can add grated parmesan, garlic or onion salt, seasoning blends, or other elements while the popcorn is warm. If you add these during popping, you run the risk of them burning while cooking.

Stove or Campfire Popcorn

Materials

- A disposable pie tin or a round tin tray from takeout, washed and dried
- Aluminum foil
- ¼ cup of unpopped popcorn
- 2 tablespoons of oil (olive, avocado, grapeseed, coconut, canola, etc.), *not* butter
- Dash of salt or more to taste
- Tongs
- Oven mitts
- A campfire or stove

Process

1. Make sure that your pan is clean and dry. Add popcorn and oil. Do not skimp on the oil. You will end up with unpopped or burned popcorn if you do. Shake the pan to make sure that the corn is covered in oil and spread out. Sprinkle with a pinch of salt.
2. Cover the pan with foil. Make sure that the foil is tightly tucked and crimped well around the edge.

3. Place the pan over heat.
    a. If cooking on a campfire or grill, make sure that the fire is steady but not raging. Place the popcorn several inches above the flames.
    b. If cooking on a stovetop, place the pan over low heat.
4. Once you hear the oil beginning to sizzle, use the tongs to grasp the pan by the side and give it few good shakes. Continue to shake the pan regularly as the popcorn cooks.
5. When the popping begins to die down, and there are about 5 seconds between pops, remove the popcorn from the heat. Let it rest a few minutes.
6. While wearing oven mitts, carefully open the foil (Figure 6.42).
7. Season your popcorn as desired—and enjoy!

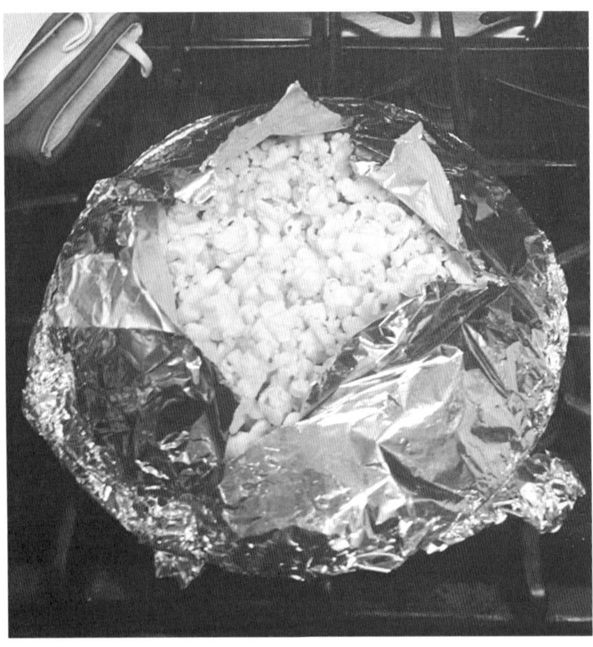

FIGURE 6.42

Microwaved Popcorn

Remember how I mentioned that microwave popcorn seems to be the popular choice these days? Well, you can make your own! It's easy, inexpensive, and healthy too.

Materials

- ¼ cup unpopped popcorn
- ¼ teaspoon of salt or to taste
- Brown paper lunch bag

Process

1. Place popcorn and salt in the bag. Shake a few times to distribute the salt.
2. Fold the top of the bag over twice to close.
3. Place the bag of popcorn in the microwave. Cook on high until there are about 5 seconds between pops (usually less than 2 minutes).
4. Open the bag carefully. Season as desired—and enjoy!

Take It Further

Because it's so easy to make popcorn, especially with the microwave method, why not "Master Chef" the experience. Provide a wide range of herbs, spices, seasoned salts, and so on, and have your campers design and taste test their own seasoning blends.

If you want to add a bit of science to your popcorn making, you can talk about why popcorn actually works. Popcorn is a special corn that has been selectively bred for thousands of years to have a higher level of moisture than other corn. When it is heated, the water and oils in the corn turn to steam. This then causes the starches inside the kernel to gelatinize, softening as they do. Soon the pressure of the steam overcomes the pliable starch and forces the kernel to burst open, cracking through the hard outer shell and immediately cooling the puffed corn starch.

Challenge campers to make the popcorn pop better. Would soaking it in water help it pop? Would drying it in the oven first change how many kernels pop? Have your campers experiment with ways to improve your popcorn-making efforts.

PROJECT 7
Sorbet Toss Game

COST: $$

MAKE TIME: 30–45 minutes

A lot of camps serve a snack during the day for campers. Provided that you have the capability to do so (check the rules of your facility and organization first), having campers make their own snack is a perfect Maker Camp activity.

The perfect way to beat the heat is to make sorbet or ice cream. There are so, so many ways to do this, but I like to keep it simple and fun. So, making sorbet becomes a game.

This project works on the simple idea that salt lowers the freezing temperature of ice. So, by adding salt to your ice mixture, you can get the temperature below 32 degrees F, helping to freeze the fruit sorbet mixture. Adding a cup of water to the salt and ice mix helps to dissolve the salt and provides better contact with the mixture in the bag, thereby making it easier to freeze your sorbet.

If you have the opportunity and ability to visit a local farmers' market, farm stand, or similar location for fresh local fruit or berries, that is a fantastic "field trip" for your campers. Give them a budget, encourage them to engage with the farmers, and let them choose the flavors for their tasty creations.

Otherwise, fruit or berries from the supermarket are fine. If you want to have the easiest possible preparation, buy frozen bagged fruit in the freezer aisle. It's already cleaned and peeled for you. Plus, if you let it partially defrost, you're already be two steps closer to a puree because the ice crystals have already broken down the cell walls and softened the fruit. Also, the cooler the fruit when you start, the faster the sorbet making goes.

Keep in mind that fruits such as melons and berries usually break down easily for sorbet, whereas more fibrous fruits such as mangos may be a little tougher to work with. You can also add a bit of juice to your mixture; just be prepared to spend a little longer working to freeze it.

I don't add additional sugar to my sorbet. Usually, I simply add a few pieces of ripe bananas instead. This adds lots of sweetness and great texture to the sorbet. However, depending on your taste preferences, the freshness of your fruit, and the fruits selected, it can help to have some table sugar or, even better, honey available.

You will need to puree your fruit before making the sorbet. If you have access to a small smoothie blender, that works very well. A stick blender and a glass jar work too. If all else fails, put the fruit into a zip-top bag, press out the air, and seal it well. Then use your hands to mash the fruit into a puree. (Many campers agree this is the most fun way to prepare the fruit.)

I usually have campers work with a partner or in a small team, just to keep the chaos slightly more organized. Encourage teams to create different flavors so that you can have a tasting and compare notes. There is nothing better than enjoying the last minutes at camp outside sharing sorbet with your camp buddies.

Materials

- 1 zip-top gallon freezer bag per team
- 1 zip-top quart freezer bag per team
- 1 cup rock salt per team
- 2 cups ice per team
- 1 cup water per team
- Roughly 2 cups fresh or frozen and defrosted fruit or berries per team
- Additional sugar or honey if desired

- Cutting boards
- Vegetable peelers
- Kitchen knives
- Sieves (optional, to remove seeds)
- Measuring cups
- Cloth towel

Process

1. Wash and dry all fruit and berries. Remove any peels, pits, stems, and seeds. Each team will need approximately 2 cups.
2. Using your preferred method, puree the fruit (see above). If you are using berries, you may want to pass the puree through a sieve to remove seeds. If desired, add sweetener and mix well.
3. Place the puree in a quart-sized zip-top bag. Gently and carefully squeeze out all the air. Make sure that the bag is sealed well.
4. Place the quart-size bag into the gallon bag. Add 1 cup of rock salt, 1 cup of water, and 2 cups of ice to the gallon bag. Squeeze out the air and seal the bag well.
5. Take turns gently tossing the bag from person to person. I suggest having campers sit in circles, tossing the bag from one person to the next. This is a great opportunity to break out a camp song such as "Down by the Banks" or create a version of "musical chairs."
6. The bag will get quite cold, so make sure that no one holds on for too long. If needed, add ice to the bag.
7. After 5–10 minutes of tossing, remove the quart bag, wipe it off with a cloth towel, and check the mixture inside to see if it has hardened to a soft sorbet. If not, continue the game (Figure 6.43).

FIGURE 6.43

Take It Further

Homemade ice cream recipes work well with the same method, although you may have to contend with allergy issues.

If campers are interested in learning more about the science behind salt and ice, consider trying your own version of the famous MythBusters Episode 29: *Cooling a Six Pack* in which Adam and Jamie use foam coolers, ice, water, and salt to determine how best to cool a six pack of beer (cans of soda are easily substituted).

CHAPTER 7

The Great Outdoors

As I said in Chapter 1, you want to "keep it campy," and that means getting outside. Even if it just means blocking off a parking lot with plastic cones so that campers can get some sunshine, taking time outside should be part of every Maker Camp, if possible. The activities in this chapter are meant to take your campers outdoors.

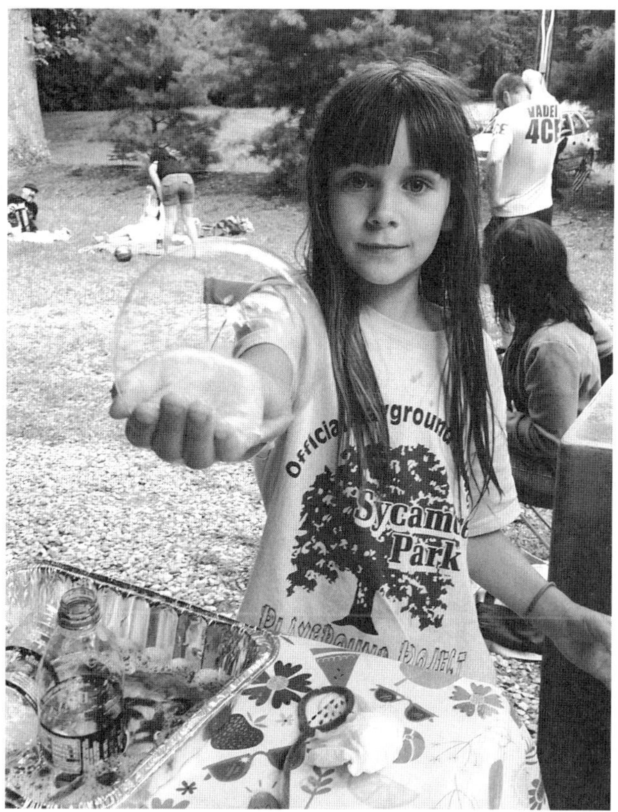

PROJECT 1
Weather Stations

There are some branches of science that kids fall in love with early on—things like mineralogy, astronomy, and meteorology. These types of science are accessible, observable, and all around us every day. Paired with children's natural desire to explore their world, it's pretty understandable that many kids create rock collections, memorize constellations, and want to understand the weather.

Creating a Maker Camp weather station can be a fun low-tech DIY activity or an exciting high-tech one depending on your budget and location. Regardless, the goal is always the same: to watch for trends over time. The ability to analyze data, draw conclusions from the data, and make predictions is important in science, technology, engineering, and mathematics (STEM). A weather station is a great real-world way to put those skills to use.

Low-Tech Weather Station Ideas

COST: Free–$

MAKE TIME: 15–20 minutes each

Generally, you'll want your weather station to be able to collect certain data. The most typical are temperature, air pressure, precipitation, wind speed, and wind direction. So you will need suitable equipment to measure these natural phenomena, specifically a thermometer, barometer, rain gauge, anemometer, and wind vane.

If your goal is to focus on hands-on making, each of these pieces of equipment provides ample opportunities for engineering and experimentation. You may want to break up your campers into teams and have each focus on a specific piece of equipment. This has the added benefit of stressing the collaborative nature of science and maker communities.

The ideas presented here are fairly simple. You can build more sophisticated equipment, of course. If you plan to use your weather station throughout the summer or during the upcoming school year, you may want to consider upgrading. Some ideas are given at the end of this section. Of course, if you are more interested in the data-collection aspect, it's okay to buy commercially available equipment and put it to good use. Many local home or hardware stores have options available.

Make sure that you have set times during the day for campers to take measurements. Divide up the work, and create a schedule. Keep a common journal where everyone can record their observations. Be sure to set aside time at the end of your data collection to analyze the data and reflect on what you've learned.

Rain Gauge

Rain is one of the most obvious weather patterns you can track. Considering how miserable a rainy day can be in summer, turning a drizzly day into an experiment can be a great experience for campers. When their rain gauge is paired with other equipment, such as a barometer and a thermometer, campers can start to see how these forces work together to cause rain and storms.

Materials

- Clean, clear 2-liter soda bottle
- Permanent markers
- Ruler or tape measure
- Scissors
- Masking tape (optional)
- Pebbles, rocks, or marbles as weights

Process

1. Carefully mark the soda bottle 2–3 inches below the neck in at least four spots using the ruler. Connect the dots with the permanent marker to make a ring around the bottle.
2. Use the scissors to remove the top. Save the top. Use masking tape to trim the raw top of the bottle, if desired.
3. Place the pebbles or marbles into the bottom of the bottle, filling the base of the bottle and creating a level surface.
4. Using the ruler and the permanent marker, measure from the top of the pebbles to the top of the soda bottle. Make marks at least every inch. You may want to make marks every ¼ or ½ inch. To make it easier to read, you may want to mark masking tape and stick it to the bottle.
5. Invert the top of the bottle, and use as a funnel at the top of the bottle.
6. Add water until it covers the pebbles and reaches the bottom of the bottom of your first mark.
7. Place the rain gauge outside in an open, uncovered location.
8. After any rain, log the number of inches of rain collected (Figure 7.1).

FIGURE 7.1

Thermometer

You can, of course, use a store-bought thermometer, but campers may enjoy making their own. It is fun to make a thermometer, and the activity opens the door to great discussions about the physical science behind how a thermometer works. his thermometer won't provide scientifically accurate measurements but will show how the process works.

Materials

- Rubbing alcohol
- Water
- Cylinder-shaped clear jar or bottle (bottles with a narrow neck work best)
- A drinking straw
- Modeling clay
- Food coloring
- Awl or screwdriver
- Safety gloves, if desired

Process

1. Remove any labels or glue from the jar or bottle.
2. Fill the bottle about one-quarter of the way with a mixture of half alcohol and half water. Add a few drops of food coloring.
3. Using the awl or screwdriver, put a hole in the jar top. Make the hole wide enough for the straw to fit through.
4. Push the straw through the hole and into the jar. The bottom of the straw should be under the liquid, near the bottom of the jar, but not actually touching the bottom.
5. Attach the top to the jar tightly. Use modeling clay to create a seal around the straw.
6. Make a note of the level of the liquid in the straw (Figure 7.2).

FIGURE 7.2

7. Place the thermometer outside in the sun. What happens? Place the thermometer in the shade. What happens?
8. (Optional) To calibrate your thermometer, place it against a wall with a piece of blank paper behind it. Mark the level of the liquid in the straw on the paper. Use a store-bought thermometer to determine the actual temperature and write that next to

the line. Repeat over the course of several days to create a scale for your thermometer.

Barometer

Air pressure is constantly changing to some degree, responding to temperature and humidity. Drops in air pressure are linked to oncoming storms, whereas high pressure is linked to more mild weather. Like the thermometer earlier, this barometer won't give you specific or accurate measurements, but it will let you track trends in air pressure.

Materials

- A wide-mouth glass jar
- A large, round balloon, any color
- A strong rubber band, string, or duct tape
- One or two drinking straws
- Glue or transparent tape
- An index card or sheet of paper
- Pen or marker
- A thermometer

Process

1. Cut the neck from the balloon; leave the rounded portion.
2. Stretch the balloon over the mouth of the jar, making it tight and taught.
3. Use the rubber bands or tape to secure the balloon well to the neck of the jar.
4. Clip one end of the straw to create a pointed shape. If you have the space, slip the end of a second straw into the blunt end of the first, and tape them together. The longer the arm, the more accurate the barometer will be.
5. Lay the other end of the straw horizontally against the balloon covering the mouth of the jar with the end in the center of the circle. Glue or tape it into place. Allow the glue to dry completely (Figure 7.3).

FIGURE 7.3

6. Place the barometer against a wall indoors. Hang the paper or index card behind the pointed tip of the straw.
7. Mark the location of the tip of the straw.
8. At least once a day, mark the location of the tip of the straw on the paper, recording the date, time, and temperature. If desired, use the National Weather Service website (weather.gov) to check the atmospheric pressure to calibrate your barometer.
9. Because the jar is sealed, no additional air can enter. This means that when the air pressure rises, the air outside the jar will press down on the balloon, and the straw arm will point upward. When the pressure drops, the arm will drop as well. Be sure to check the thermometer to ensure that changes in the position of the barometer arm aren't the result of shifts in temperature.

Anemometer

An anemometer is a device that is used to measure wind speed. This simple cup version is a classic that catches the wind and causes the device to spin. Campers will need to be present during the wind to count the number of rotations over time to calculate the wind speed. If possible, pick a windy day or use an electric fan to simulate one.

Materials

- 5 small paper cups
- 2 plastic drinking straws
- Tape
- Scissors
- Hole punch
- Straight pin or pushpin
- Pencil with a new eraser
- Stapler
- Permanent marker
- Stopwatch

Process

1. Take one cup and place four holes near the rim equidistant from one another at quarters. Place a hole in the center of the bottom of the cup.
2. Feed the straws through the holes, letting them cross in the center. Tape the straws together in the center where they make a "T."
3. Push the pencil up through the hole in the bottom of the cup, leading with the eraser. Use the pin to secure the straws to the eraser. Make sure that the straws can spin freely. Wiggle the pin to widen the hole if needed.
4. Place the other cups at the ends of the straws with the straw on top. Make sure that they are level and straight. Staple or tape them in place. All the cups must face in the same direction (Figure 7.4).

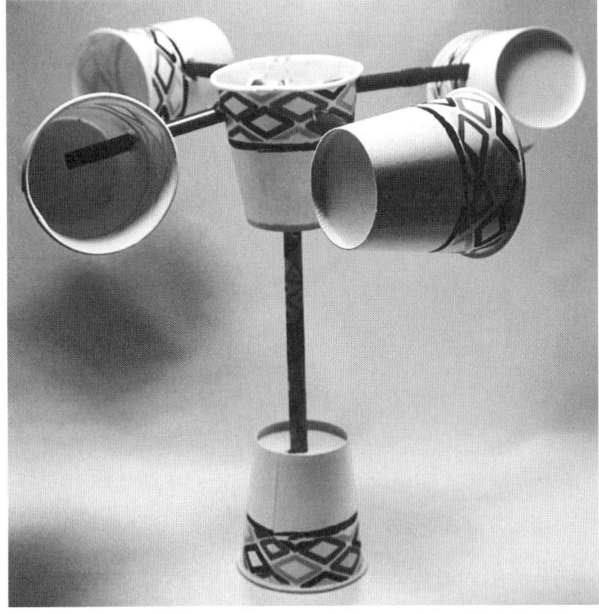

FIGURE 7.4

5. Mark one cup with an X using permanent marker.
6. To use, place the anemometer in the wind. Decide on an amount of time to measure the rotations of the anemometer, such as 30 seconds. Using the marked cup as a reference, count the number of times the cups spin over the course of that time. Divide the number of rotations by the time.
7. Technically, to calculate the speed of the wind, you need to calculate the actual distance one of the cups is traveling. To do this, you need to calculate the circumference of the circle it travels. Measure the distance from one cup, across the center cup, to the cup opposite. This is the diameter d. The

circumference C can be calculated using the formula $C = d\pi$, where π is roughly equal to 3.14. Multiply this circumference by the number of rotations. Then divide this number by the time. Using unit conversion, you can transform this number into miles per hours and compare it with the wind speed reports by local weather websites.

Wind Vane

A wind vane, also known as a *weathervane*, is a classic piece of weather equipment. This simple device points the direction of the wind. If you have a particularly creative group, let them use their artistic skills to make the wind vane beautiful.

Materials

- A plastic or paper cup
- A paper plate
- A plastic straw
- Straight pin or pushpin
- Pencil with a new eraser
- An index card
- Scissors
- Ruler
- A pencil or marker
- Tape or hot glue
- Awl or screwdriver
- Compass or compass app on a smartphone

Process

1. Using the ruler, divide the plate into quadrants. Mark them "North," "East," "South," and "West."
2. Place the cup on top of the plate in the center, with the mouth of the cup facing down against the plate. Tape or glue the cup into place.
3. Use the awl or screwdriver to make a hole in the center of the bottom of the cup.
4. Push the pencil into the cup with the eraser facing up.
5. Cut an arrow and square from the index card. Cut slits in the ends of the straw, and slide the arrow into one end and the square into the other. Use tape or glue to secure the card pieces to the straw. Make sure that the arrow and square are straight.
6. Place the center of the straw against the pencil eraser. Push the pin through the straw and into the eraser. Make sure that the straw is level and can move freely.
7. Take the weathervane outside. Use a compass to determine north, and orientate your weathervane to match. Use tape or rocks to secure the plate to a hard surface. Wait for the wind, and track the direction (Figure 7.5).

FIGURE 7.5

High-Tech Weather Station Ideas

COST: $$–$$$

MAKE TIME: 60–90 minutes each

If you are looking for a weather station that provides more accuracy, you have many options. However, they will require digital equipment, a larger budget, and a more permanent location. There are a number of kits available that either provide everything you need or provide equipment that will work with existing microcontrollers that you can code. We'll look at an example of each.

FIGURE 7.6

The Complete Kit Option

Kits such as those made by Acurite (www.acurite.com/) or Davis (www.davisinstruments.com/) include everything you need to set up and start a weather station quickly and easily. They are generally reliable and provide easy-to-use interfaces to track the weather. Although the price tag can be high, look into discounts on older models or educational discounts if you work with a school (Figures 7.6 and 7.7).

If you are using this kind of kit, check to see if it includes software that will let you download data. You can use this software to create a free Weather Underground personal weather station (www.wunderground.com/weatherstation/overview.asp). Community outreach and communication are key maker and science skills, so creating and maintaining a weather station for the summer can be tremendously empowering.

If you have a green screen available, consider having the campers plan a weather recap video based on the data they've collected. Creating backdrops, graphics, animations, and a solid script are fantastic science, technology, engineering, arts, and mathematics (STEAM) activities.

FIGURE 7.7

Building a Weather Station with an Open-Source Controller

Makers have created weather stations using a variety of microcontrollers, such as Adrunio and Raspberry Pi. By combining the controller with various sensors and customized code, one can create a sophisticated device for measuring the weather. Examples of these truly DIY efforts include

- Gareth Branwyn's Arduino-Based Weather Station for Make Media (https://makezine.com/2015/11/20/build-your-own-arduino-weather-station/).
- Marc-Olivier Schwartz's WiFi-enabled Arduino Weather Station from Adafruit (https://learn.adafruit.com/wifi-weather-station-arduino-cc3000/introduction).
- In 2016, the Raspberry Pi Foundation launched the Raspberry Pi Oracle Weather Station kit with a Hardware Attached on Top ("HAT") add-on board that fits onto the Raspberry Pi to extend its functionality. Now the foundation provides the complete DIY instructions for educators on its website (https://projects.raspberrypi.org/en/projects/build-your-own-weather-station).

However, if you're looking for a good balance of cost-effective DIY aspects and ease of setup, SparkFun's weather:bit kit for the popular micro:bit is a great option. It includes all the equipment and sensors you need in an easy-to-build kit that monitors temperature, humidity, air pressure, wind speed, and wind direction as well as soil temperature and moisture, which are nice additions if you have a school garden (Figure 7.8).

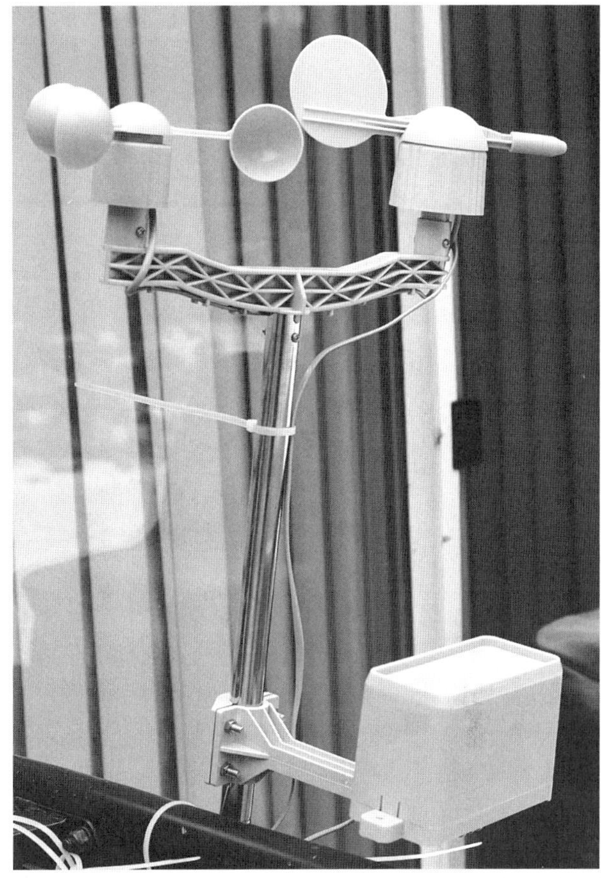

FIGURE 7.8

Code your micro:bit using MS MakeCode, and simply slide it into the weather:bit. Although it doesn't transmit the information over WiFi, the micro:bit itself does have both radio and Bluetooth capabilities, and the weather:bit includes a data logger with an SD chip so that you can retrieve the information and analyze it in a spreadsheet program such as Excel (Figure 7.9).

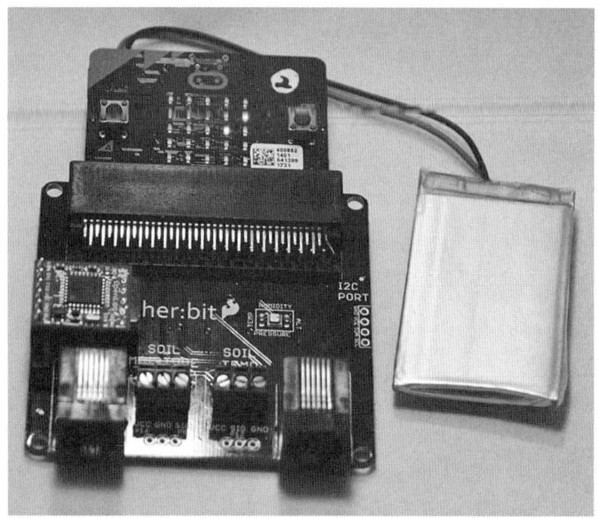

FIGURE 7.9

The weather:bit is easy to code with MS MakeCode, which gives campers the opportunity to adapt the code to collect the data they want. For example, I adapted the base code to allow me to set the time on the micro:bit so that our logged data were easier to read. I also added the ability to do a spot check of weather conditions with the push of a button so that the micro:bit gives a real-time readout of temperature, humidity, wind speed, and wind direction (Figure 7.10). You may decide to add features as well. A complete copy of my code is available at https://makecode.microbit.org/_gCiff5XMWFJt.

The kit doesn't come with a waterproof enclosure for the weather:bit, micro:bit, and power source, so you will need to design one to keep your electronics safe. I've used a modified Rubbermaid container with success. Full documentation is available on the SparkFun website (https://learn.sparkfun.com/tutorials/microclimate-kit-experiment-guide/about-the-weatherbit).

To connect your micro:bit to computer or smartphone, I suggest the apps offered by Bitty Software (https://www.bittysoftware.com/). Using the Bluetooth on the micro:bit., these apps offer the ability to log data in real time, graph the results, and set alerts. Older campers may enjoy customizing the code to track your weather.

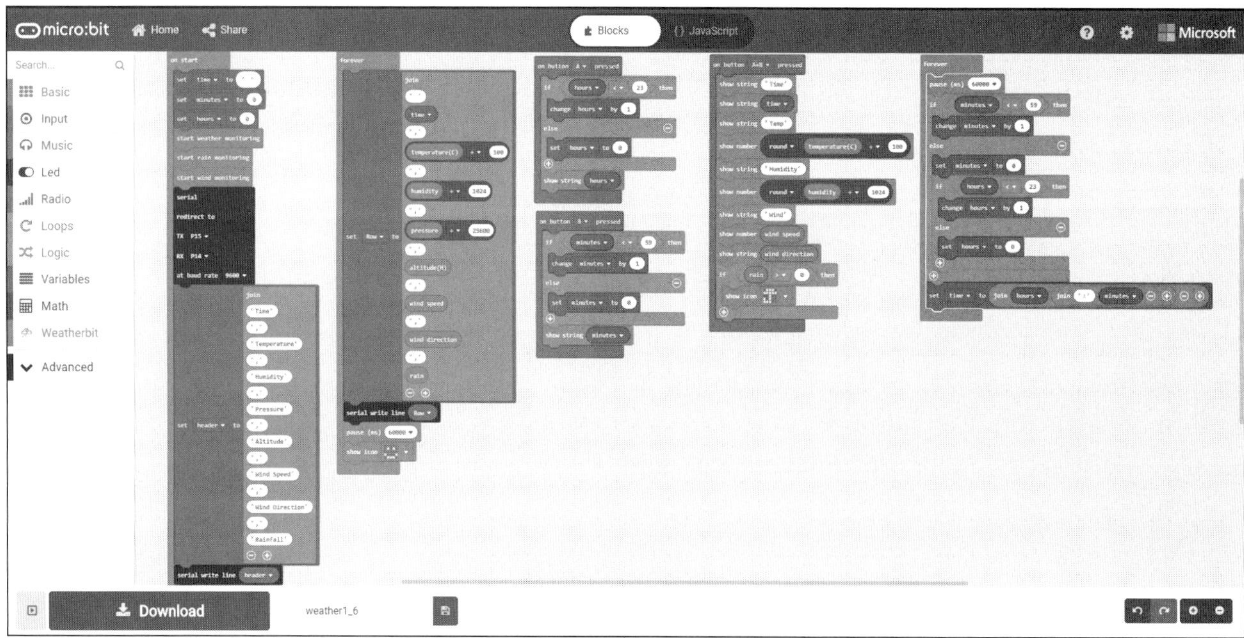

FIGURE 7.10

PROJECT 2
Citizen Science at Camp

We live in amazing times. With technology and the Internet, there is an unprecedented opportunity for anyone to become part of the scientific process. Now scientists across the globe are creating citizen scientist initiatives that put data collection and more in the hands of the average person.

As makers, our campers tend to be attracted to new things and exciting discoveries. The citizen science programs listed below are fantastic ways for campers to explore their world while at Maker Camp. This listing features just a small number of the projects available, so if you don't see something that sparks your interest, go online and head to SciStarter (https://scistarter.com/) for many additional options.

iNaturalist

One of the largest communities of naturalists and citizen scientists online, iNaturalist, has a very simple goal: to empower citizens to record biodiversity. To participate, all you need is a camera of some kind, access to the internet, and a willingness to share what you observe, making it an easy fit for Maker Camp. Taking and sharing photos of local plants and animals can quickly become addictive. The website also hosts a tremendous amount of information on various species, so campers can identify their findings while providing important information to scientists. You can also easily view recent observations made by others in your community, making the entire project highly collaborative. For educators, there are tools to create local projects and BioBlitzes that can enhance your Maker Camp. Learn more at www.inaturalist.org/.

GLOBE Cloud

This year-round project is hosted by NASA as part of their GLOBE Observer citizen scientist offerings. It compares human-recorded cloud data with NASA satellite data to collect data on cloud type, height, cover, and related conditions from all over the world. Because clouds play a significant role in the temperature and energy balance of the Earth, their study is vital as scientists work to understand the systems of our planet. Class and training materials are available for free for all ages. Visit https://observer.globe.gov/do-globe-observer/clouds to learn more about participation.

Journey North

Founded in 1994, Journey North is one of the oldest and most established citizen science programs in North America. Through a variety of projects, the organization aims to track the migration of different species across the continent and across the seasons. Whether your campers are interested in birds, frogs, worms, or whales, there is a project sure to capture their interest. Citizen scientists use the website to track sightings of specific species. It's as easy as that! A wide variety of educator resources is available, so there are many opportunities to learn and share. Find out more at https://journeynorth.org/.

Nature's Notebook

The Nature's Notebook program has amateur and professional naturalists observe plants and animals in their local areas, record what they see, and report their findings to the National Phenology Network to create long-term data sets used for scientific discovery. The program provides a wide range of species you can choose to follow, so you may want to limit choices with a little research before your Maker Camp begins. You'll also need to set up an account ahead of camp. Campers can collect data on paper or online using the app, so it's easy to participate regardless of your technology level. Ideally, this project is suited to camps that will meet several times over the course of the summer. To learn more, visit www.usanpn.org/natures_notebook.

Urban Buzz

The Urban Buzz cicada collection program is one of the easiest citizen scientist programs to incorporate into your Maker Camp. All you need to do is go out and find between two and five cicadas, identify your habitat type, make some notes about the bugs, record observations about the day, and mail the insects to the scientists. Because cicadas are sensitive to changes in their environment, especially temperature and the availability of trees, researchers track them to understand how human changes to the environment, specifically urbanization, may affect other species. Learn more at http://studentsdiscover.org/lesson/urban-buzz-citizen-science-with-cicadas/. While you're there, check out other Student Discover projects led by scientists and researchers.

Mosquito Byte

Here is another easy way to participate in citizen science. Simply download an app and track when you get bitten by mosquitos. The data are compiled by scientists to produce a global map of mosquito activity. By tracking your data with other weather information, scientists are working to track potential pathogens and infection outbreaks. Learn more at http://vectorecology.org/outreach/mosquito-bite-app/.

Geo-Wiki Picture Pile

This citizen science project is great for downtimes and lulls that happen as campers finish projects or gather to start new ones. The scientist at Geo-Wiki are working to monitor environmental changes worldwide. The "Picture Pile" game, available online and as an app, asks citizen scientists to sort pictures of everything from clouds, to forests, to human light patterns at night. By crowd-sourcing the work, researchers are better able to identify environmental changes and problem areas. Learn more at https://geo-wiki.org/games/picturepile/.

PROJECT 3
PVC Stomp Rocket Launcher

COST: $

MAKE TIME: 20–30 minutes for the launcher, 10–20 minutes for the paper rockets

This rocket launcher, based on a design on NASA's education site, has been a hit at every party, camp, or event I have ever brought it to. You can carry most of the parts, unassembled, in a 1-gallon zip-top bag, so it's easy to transport anywhere. It goes together quickly, and only costs about $5 to make, so you can have several ready to go at any time. The rockets are made of paper, so they're inexpensive as well, giving campers many opportunities to prototype and experiment with their designs (Figure 7.11).

Materials

- 1 empty (and rinsed) 2-liter plastic soft drink bottle
- 2½-inch PVC tee connectors
- 1½-inch PVC connector
- 2½-inch PVC caps
- 5-foot length of ½-inch PVC pipe
- Duct tape
- Hacksaw or PVC cutter
- Copies of the rocket template on cardstock

FIGURE 7.11

- Scissors
- Invisible tape
- A piece of napkin, paper towel, or scrap paper
- A wooden dowel or a spare piece of PVC
- Markers and stickers to decorate
- Goggles

Process: The Launcher

1. Cut the PVC into 3- × 12-inch-long pieces and 3- × 6-inch long pieces.
2. Assemble the pieces as shown in Figure 7.12.
3. Tape the soda bottle to the end of the launcher, as shown, using duct tape.

Process: The Rocket

1. Print copies of the template onto cardstock (Figure 7.13). If printing to regular printer paper, paste the copy to stiffer materials, such as manila folders.
2. Color and decorate your rocket as desired.
3. Cut along the solid lines.
4. Once cut, fold along the dotted lines to create a three-sided "tube" for the rocket. Tape along all the edges, being sure not to leave any gaps through which air can escape.
5. Fold the three pieces of the nose together, and tape well. Don't leave any gaps.
6. Fold the fins outward.

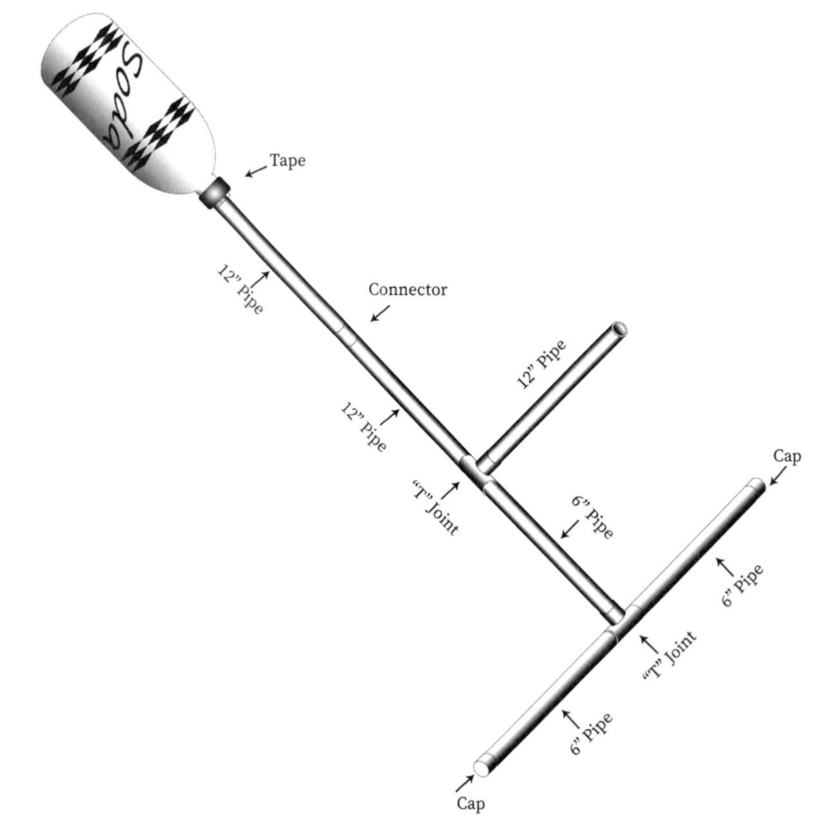

FIGURE 7.12 Diagram by Jennifer McBride

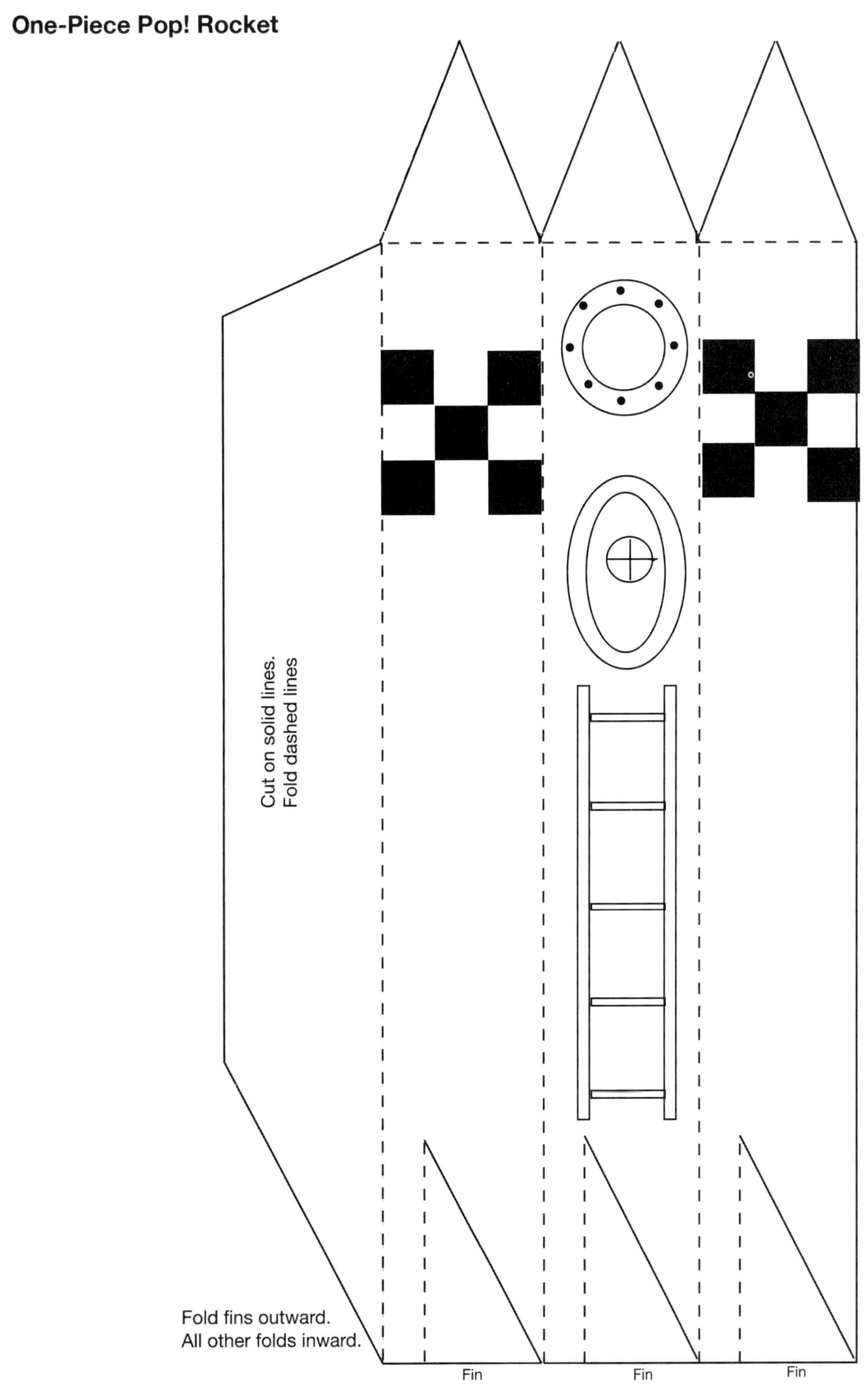
One-Piece Pop! Rocket
Cut on solid lines.
Fold dashed lines
Fold fins outward.
All other folds inward.
Fin
Fin
Fin

FIGURE 7.13 Template by Jennifer McBride

7. Using the wooden dowel or spare PVC pipe, push some napkin into the nose to add weight for steadier flying. (You can adjust this ballast as needed to improve flight.)

Process: Launching

1. Place the launcher in an open space, and tilt the launch tube in the desired direction. If there is a light wind, aim in the direction of the wind. Make sure that the landing zone is clear of anyone who might be hit by the rocket.
2. Place the paper rocket in the launcher. Do not push it too far down the tube. Less friction is a good thing.
3. Put on eye protection, and do a countdown to zero.
4. Stomp or jump on the label of the bottle. This will force most of the air inside the bottle through the tubes and launch the rocket.
5. To reinflate the 2-liter bottle, separate the bottle from the launcher by pulling it from the connector. Wrap your hand around the pipe end to make a loose fist, and blow through the opening into the pipe. Doing so keeps your lips from touching the pipe. Reconnect the bottle to the launcher, and it is ready to go again.

PROJECT 4
Bubbles

When the weather turns warm, my campers and I love to break out the soap and make bubbles. There are so many scientific principles behind the little balls of fun as well. They're great for play, art, and more. Of course, when it comes to bubbles, I tend to feel that bigger is better, so I'm going to share some ways to take your bubble making to the extreme!

A bubble is just air wrapped in soap film. Soap film is made from soap and water (or other liquid). The outside and inside surfaces of a bubble consist of soap molecules. A thin layer of water lies between the two layers of soap molecules, sort of like a water sandwich with soap molecules for bread. They work together to hold air inside.

Why is a bubble round? Bubbles can stretch and become all kinds of crazy-looking shapes. But, if you seal a bubble by flipping it off your wand, the tension in the bubble's skin shrinks to the smallest possible shape for the volume of air it contains. This is why even if it had a goofy shape before you sealed it, once sealed shut, the bubble will shrink into a sphere shape. Compared with any other shape, a sphere has the smallest surface area versus volume.

A bubble gets its color from light waves reflecting between the soap film's outer and inner surfaces. The distance between the layers gets smaller as the water evaporates, making the colors change. Bubbles can also reflect what's around them, such as the faces peering at them.

For Maker Camp, get outside and make some bubbles. Once you have made your mix, there are so many things you can do.

Making Bubble Mix

COST: Free–$

MAKE TIME: 5–10 minutes to make the mix, overnight to rest

Materials

- 2 cups regular dish detergent (not the antimicrobial kind)
- 1 gallon distilled water
- 1 tablespoon vegetable glycerin

Process

Mix all the above together, stirring gently. If possible, let it sit overnight before using. If it is kept covered and clean, the bubble mix can be used and reused for weeks.

Giant Bubbles

COST: $–$$

MAKE TIME: 20–30 minutes to prepare the mix

It's easy to make giant bubbles with the preceding bubble mix. These impressive creations never fail to delight campers of all ages. It's also the perfect group activity, where you can challenge campers to work together to get the biggest, longest-lasting bubble they can create.

Materials

- 3–5 gallons of bubble mix, prepared as indicated earlier and allowed to sit overnight
- A solid plastic kiddie pool, about 3 feet in diameter
- A hula hoop that fits inside the pool with a little clearance

Process

1. Place the pool on a level surface. If you are working on a smooth surface, consider adding mats to prevent the area from becoming slippery.
2. Gently add your bubble mix to the pool, taking care not to create too much foam.
3. Place the hula hoop in the pool, making sure that it is covered by bubble mix. Make sure that your hands are also coated in bubble mix.
4. Raise the hula hoop out of the bubble mix, and then sweep it to the side, creating a giant bubble. It may take a few tries to get just the right speed and movement (Figures 7.14 and 7.15)
5. If desired, let campers remove their shoes and step into the center of the pool. Holding the hula hoop around them and in the bubble mix, have them attempt to make a bubble around their bodies. Be very careful entering and exiting the pool. Do not reuse bubble mix if campers have been in the pool.

FIGURE 7.14

FIGURE 7.15

Bubble Wands

COST: Free–$

MAKE TIME: 5–10 minutes

Making your own bubble wand not only allows for exploration into the physics behind the shape of bubbles, but it's also a fun creative experience. Here I provide several ideas, each a little more challenging than the last. A day spent on inventing a better bubble wand would not be a day poorly spent (Figure 7.16).

FIGURE 7.16

Chenille Stem Wands

This is the simplest and cheapest bubble maker you could ever want. Simply provide lots of various chenille stems in colors and bowls of bubble mix. Have campers twist the stems in various shapes and sizes, and then test each with the bubble mix. You can even provide stems of various thicknesses—they are commonly sold in thicknesses of 4, 6, and 12 millimeters and larger. Or try changing the types of materials—providing both the classic fluffy versions and the metallic stems.

Bubble Blowers

Bubble blowers are easy to make. Cut the bottom off a disposable water bottle, cover it with fabric or a piece of paper towel, use a rubber band to secure, and dip it in a bowl of bubble mix. When you blow into the neck, you'll create a long "bubble snake." Try changing the cover materials, the length of bottle, or the width of the bottle to create the best blower.

Bubble Nets

COST: $

MAKE TIME: 30–45 minutes

Bubble nets made with yarn or cotton string can create many big bubbles at once. They take a little time and patience to make but are worth it, especially with older campers looking for a challenge. The simplest version is a tristring, which basically features a long string held between two long wooden dowels and a loop or triangle of string below. From there, campers can tie in additional string to create more triangles, making a net if desired (Figure 7.17).

FIGURE 7.17

As far as materials go, the most important is the string used. Many bubble enthusiasts recommend mercerized yarn. Mercerization is a process that uses a treatment with lye to make the fibers stronger, smoother, and more able to take dye. For our purposes, it helps the fiber hold the bubble mix without stretching too much while still allowing good bubbles. Use a lightweight cotton that isn't too thin.

Materials

- 2 wooden dowels, ½ or ⅝ inch in diameter, 3–4 feet
- Mercerized yarn or another cotton string, approximately 2 yards per camper
- A weight, such as a bead, washer, or small piece of PVC (optional)
- Duct tape
- Scissors
- Bubble mix in a 3- to 5-gallon bucket

Process

1. Measure a 2-foot section of yarn.
2. Wind the yarn around the top of one of the dowels several times and tie tightly. Tie the other end of the yarn to the other dowel (Figure 7.18).

FIGURE 7.18

3. Cut a piece of yarn about 4 feet long. Tie one end to the top of one dowel. If desired, tie the weight to the center of the yarn (Figure 7.19). Then tie the other end to the other dowel. You should have a wide loop or triangle (Figure 7.20).

FIGURE 7.19

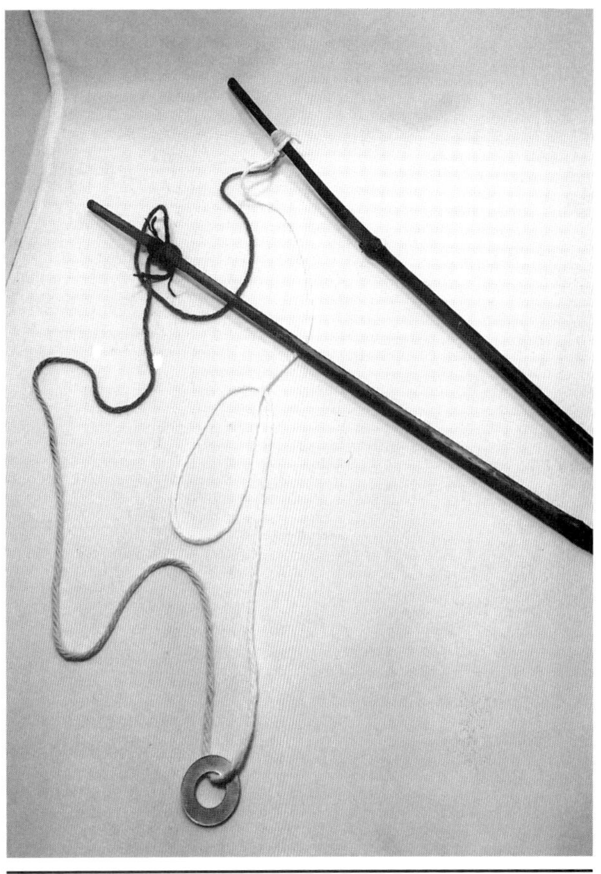

FIGURE 7.20

4. Dip the loop into the bubble mix. Lift it out, allowing excess bubble mix to drop off for a moment. Then slowly pull the bubble net through the air, creating a large, long bubble.

5. If desired, campers can use additional yarn to tie between the top string and the bottom loop, creating rectangles or triangles that will make additional bubbles. Tying yarn horizontally and knotting it at each vertical string will further divide the loop (Figure 7.21). The shapes do not need to be perfect to create many fantastic bubbles (Figure 7.22).

FIGURE 7.21

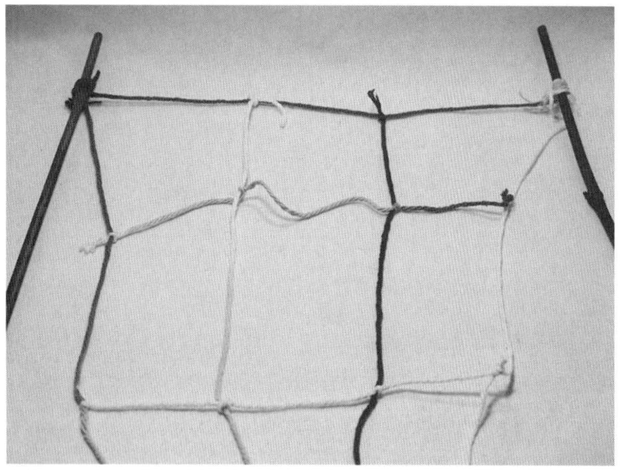

FIGURE 7.22

This project can be easily downsized to use a smaller amount of yarn and bamboo skewers or pencils if time, space, and/or resources are tight.

3D-Printed Bubble Wand

COST: $

MAKE TIME: 15–30 minutes

This is a fun way to use technology to create a fun real-world item. It's perfect for beginners because a wand is a fairly simple design to create. For a discussion on selecting 3D-printing software, please see Chapter 3. We will be using Tinkercad (www.tinkercad.com) for this project.

Process

1. Open a new project. Drag on a tube. Set the radius to 26 millimeters (1 inch) (for 52 millimeters or 2 inches in diameter). Set the wall thickness to 6 millimeters (1⁄4 inch). Set the height to 4 millimeters (1⁄8 inch). This will form the opening across which the film or bubble mix will form (Figure 7.23).
2. Drag on a box. Set the length to 100 millimeters (4 inches). Set the width to 6 millimeters (1⁄4 inch). Set the height to 4 millimeters (1⁄8 inch). Move this box so that it is centered under the circle. Make sure that it overlaps the circle a bit (Figure 7.24).

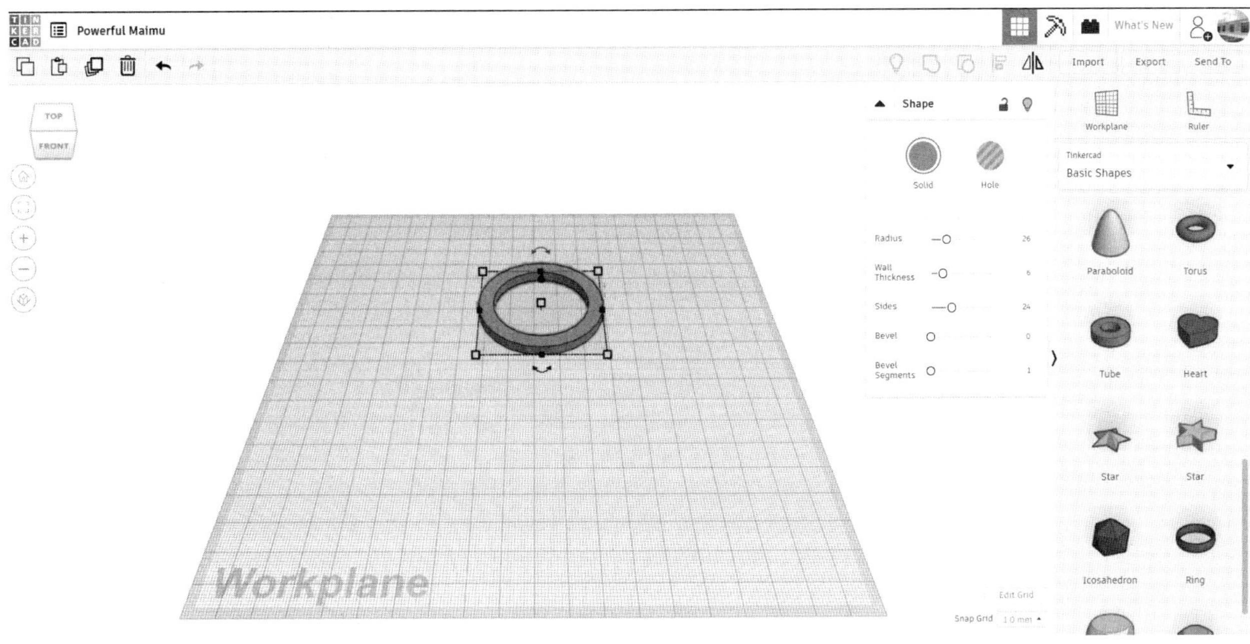

FIGURE 7.23

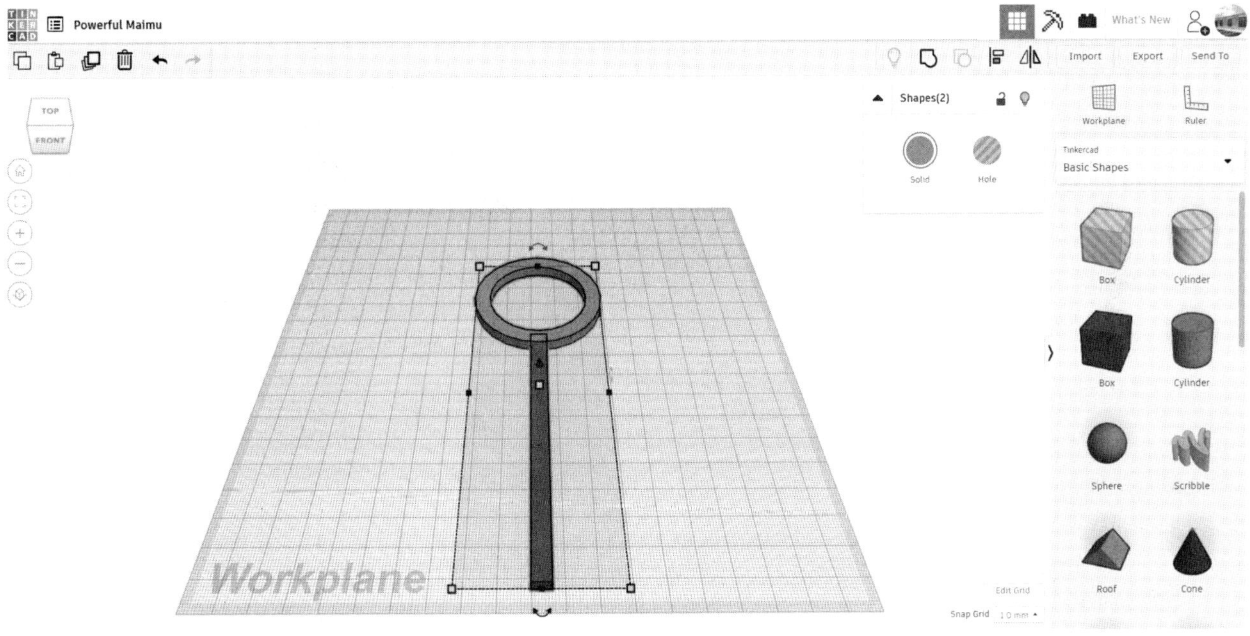

FIGURE 7.24

3. On the bottom of the box, opposite the circle, add a circle or other shape to act as a handle. Set the width to 26 millimeters (1 inch) or less. Adjust the length to roughly the same. Set the height to 4 millimeters ($^1/_8$ inch). Move this object so that it overlaps the bottom of the box slightly (Figure 7.25).
4. Select all the pieces, and group them.
5. Add wells to hold additional bubble mix for better bubble blowing (optional). To do this, create a small box that is 2 millimeters wide, 4 millimeters long, and 8 millimeters tall. Place it on the bottom of the circle. Copy and paste the box, and drag it to the top of the circle (Figure 7.26).

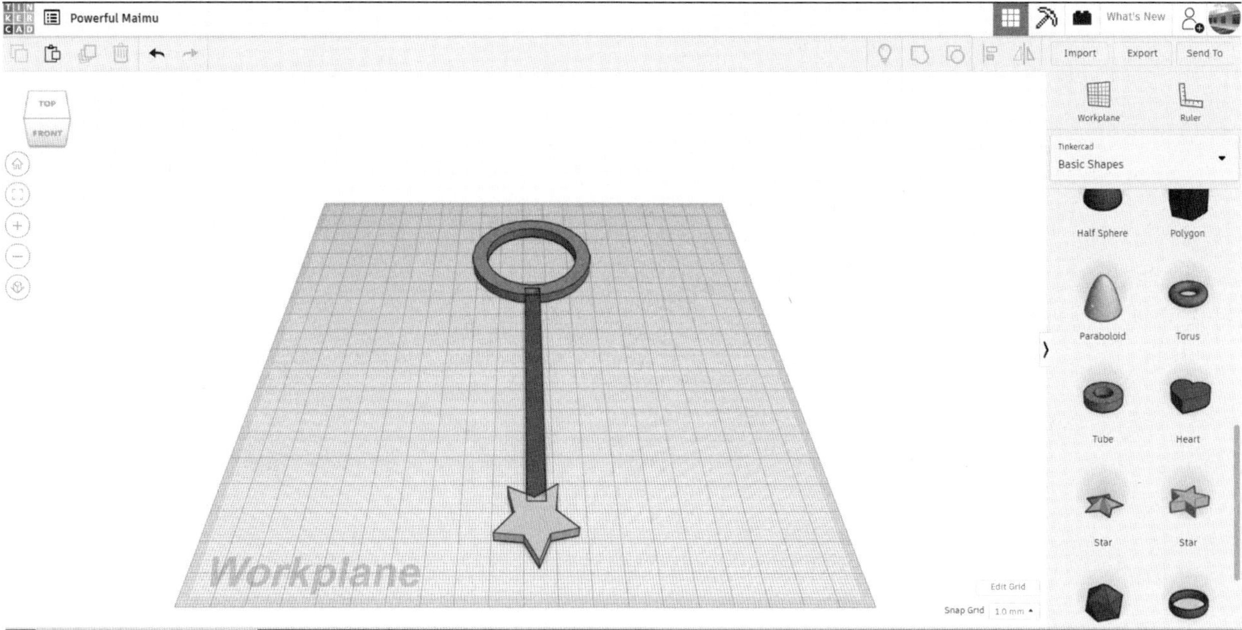

FIGURE 7.25

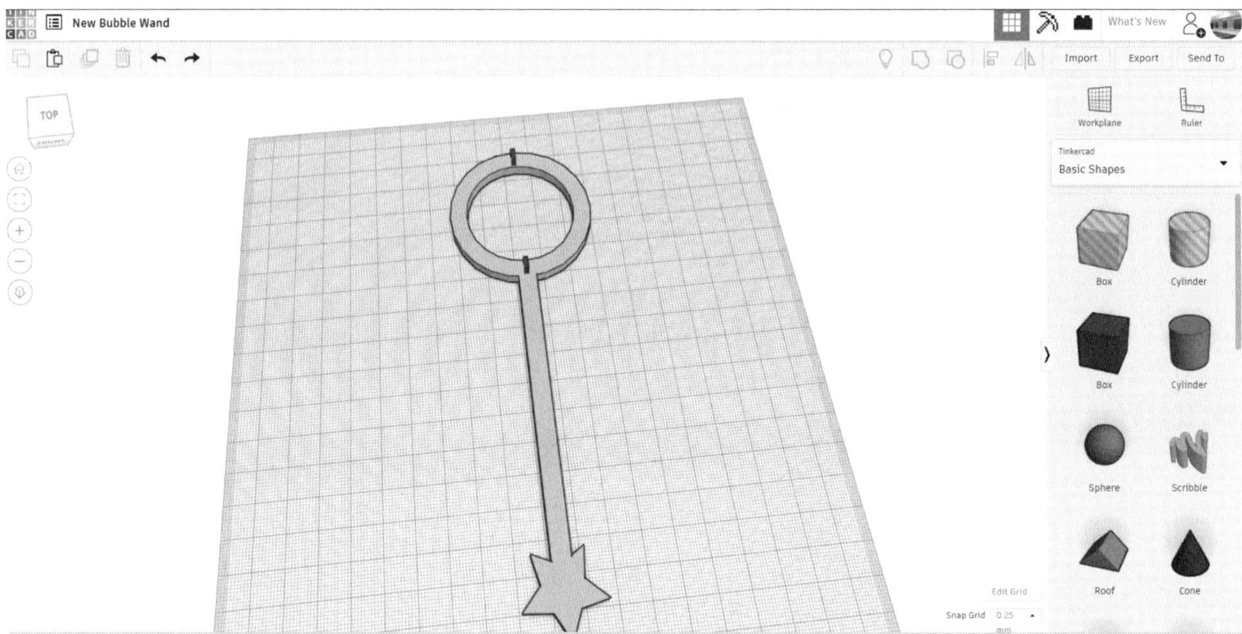

FIGURE 7.26

6. Select both small boxes at once, and use the "Duplicate" feature in the top-right corner or CTRL+D to duplicate the boxes. Select the boxes, and use the curved symbol with arrows on either end to adjust the angle by 5 degrees so that the new boxes are offset from the first ones but fit the circle (Figure 7.27). Then click "Duplicate" again. Tinkercad will automatically place the next set at 5 degrees of rotation. Continue until the circle is filled. Group the items (Figure 7.28).

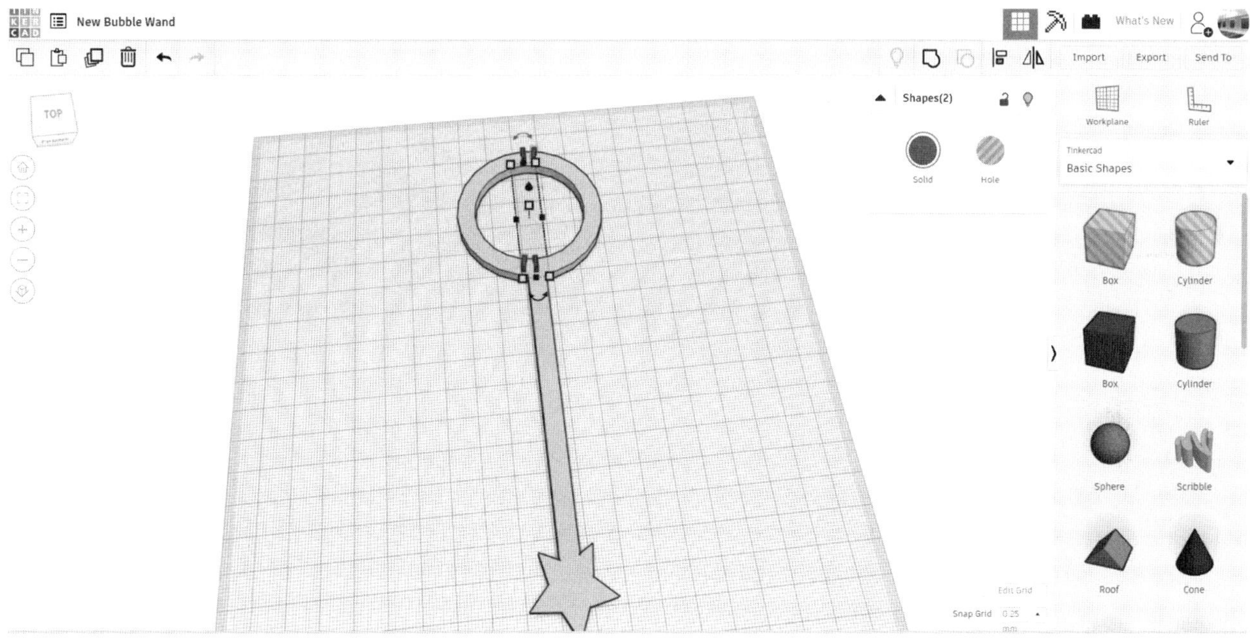

FIGURE 7.27

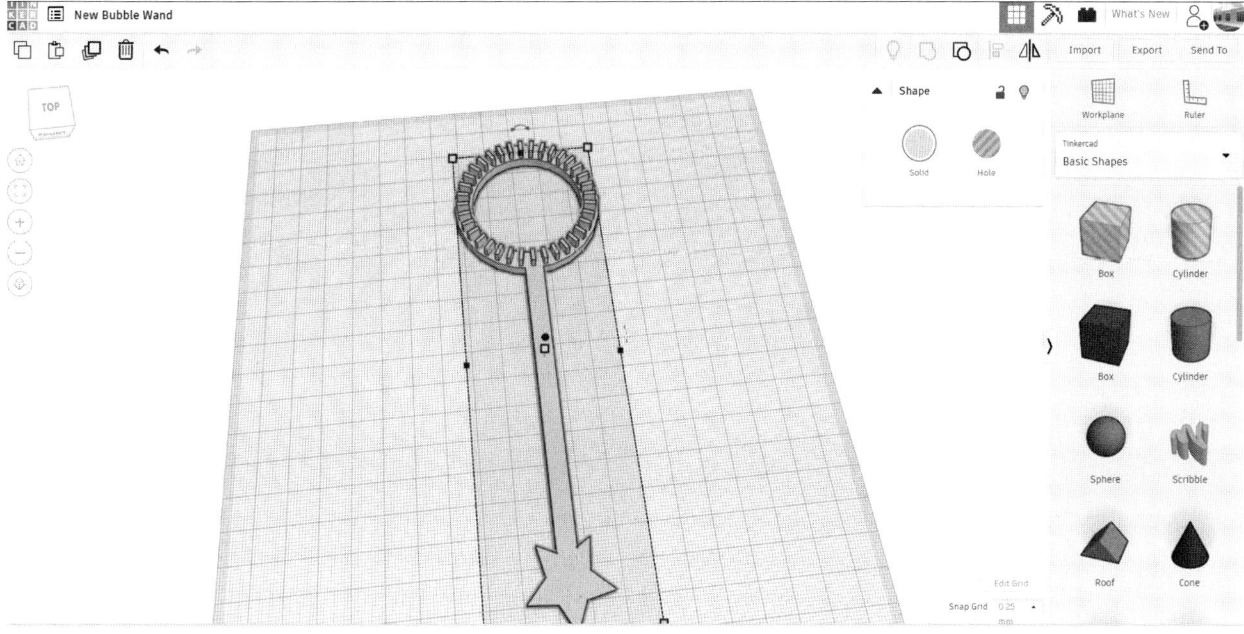

FIGURE 7.28

FIGURE 7.29

7. Export your file, and print it on your 3D printer (Figure 7.29).

NOTE: You can download this print file from my Thingiverse page at http://www.thingiverse.com/thing:3762981.

Campers can, of course, have fun with the size and shape of the opening. Encourage them to make square-, heart-, star-, or other-shaped bubble wands. Do you always get a round bubble? Why? Or try different densities of wells to see if it makes a difference when you blow bubbles.

Recently, Tinkercad added a "Codeblocks" feature to the CAD software. This allows users to drag and drop blocks of code to create their designs. Some campers may enjoy the challenge of creating an algorithm to make their bubble wands. Here's one way to do it.

Process

1. Drag over the "create new object" block. Name this object "Bubble Wand." Drag over the "tube" block. Set the radius to "26." Set the wall thickness to "6." Set the height to "4." Set the object to a solid color.
2. Add the "move" block, and set "Y" to "50." (This will move the circle toward the top of the workplace.)
3. Drag over the "box" block. Set the length to "100." Set the width to "6." Set the height to "4." Set the object to a solid color. Add the "move" block and set "Y" to "–23."
4. Add a block for the shape you'd like for your handle. Set the height to "4." Adjust the other setting to the width and length you like. Set it to a solid color. Add the "move" block, and set "Y" to "–80."
5. Drag over the "create group" block. Then add the "move" block, and set "X" to "–50" and "Z" to "2." This will move the wand out of the center workspace and lift it until it is even with the workplane.
6. Drag over the "create new object" block. Name this object "Texture." Drag over the "box" block. Set the length to "1." Set the width to "4." Set the height to "4" (Figure 7.30).
7. Drag over the "create new object" block. Name this object "Pattern." Here we will use code to duplicate the "texture" object and rotate it in a circle (Figure 7.31). Drag over the "create variable" block, and name the variable "Angle." Set this to "5." Drag in a "repeat" block. Set it to repeat 36 times.
8. Drag in the "add copy of object" block. From the "Data" menu add the "texture" variable. Set this to a solid color. Add the "rotate around" block. Set the "Axis" to "z." Add the "angle" variable from the "Data" menu. From the "Math" menu add the "X Y Z" block, and set "X" to "22," "Y" to "0," and "Z" to "0."
9. Drag in the "change" block, and set it to change angle by "5."
10. Re-create the same code as described in step 8, but instead of setting it to a solid color, select the transparent "hole" instead.

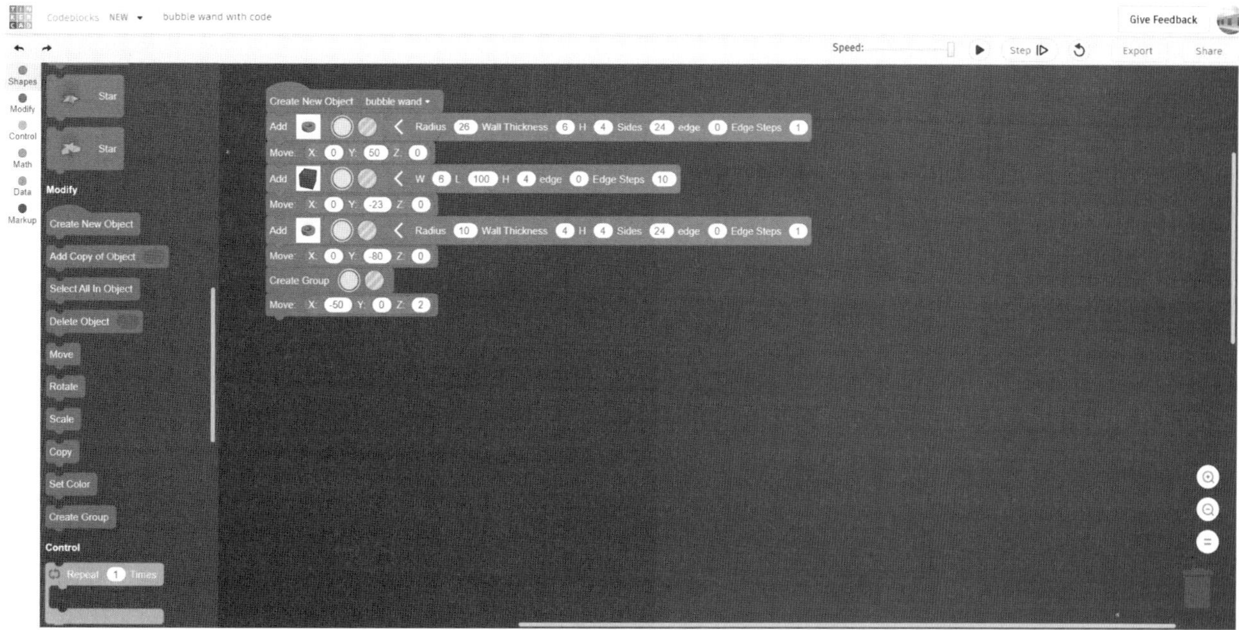

FIGURE 7.30

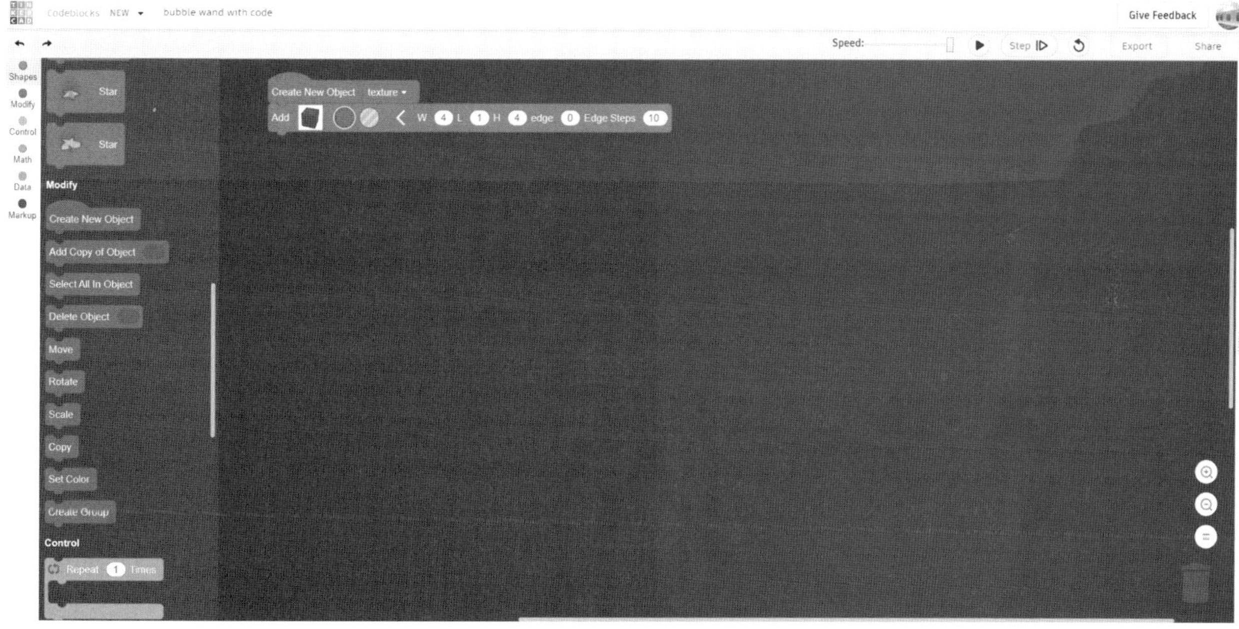

FIGURE 7.31

11. After the "repeat," drag in the "delete object" block, and add the "Texture" button.
12. Add the "create group" block, and set it to a solid color. Then add the "move" block, and set "X" to "-72," "Y" to "50," and "Z" to "4." This groups and moves the pattern.
13. Drag in the "select all in object" block. Then add "create group," and set it to your desired color (Figure 7.32).
14. Click the "Run" arrow to create your object, and debug as needed (Figure 7.33).

NOTE: A complete copy of the code is available at https://www.tinkercad.com/codeblocks/1qH9dtFgr0q.

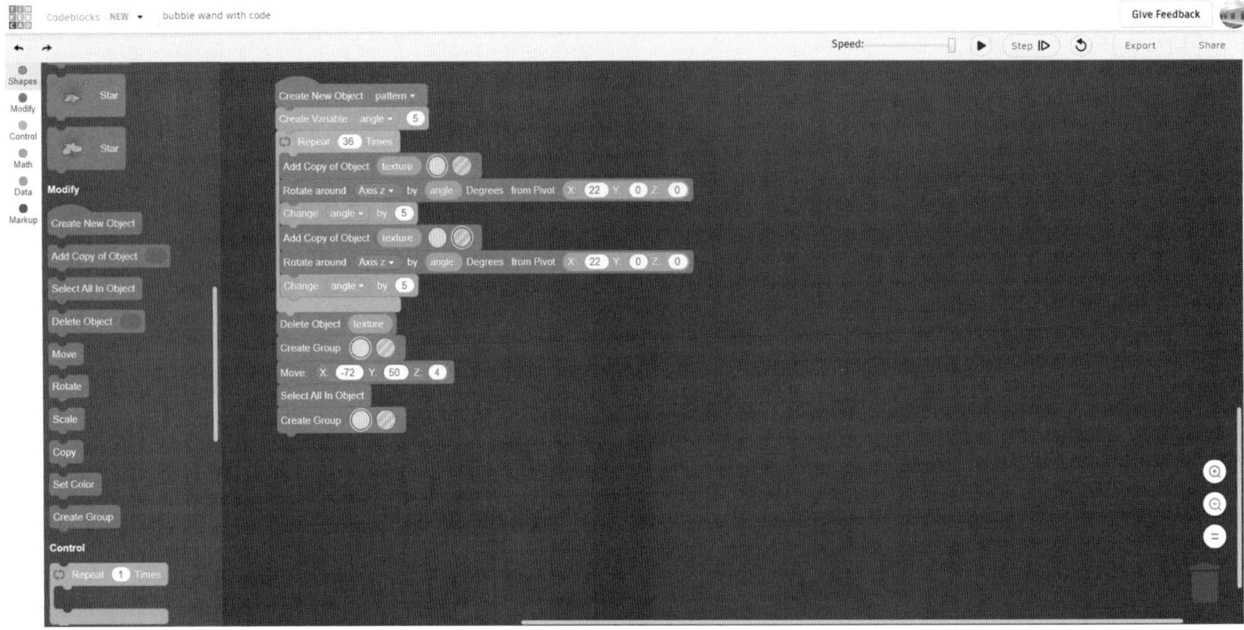

FIGURE 7.32

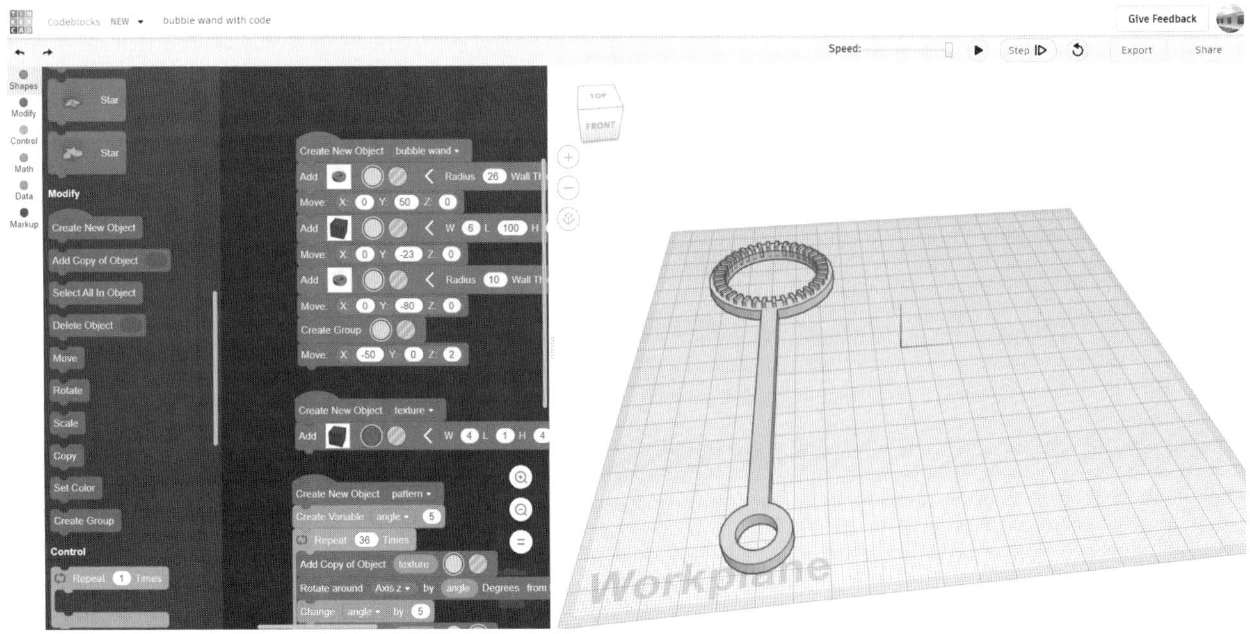

FIGURE 7.33

Bubble Prints

COST: $

MAKE TIME: 10 minutes

As a fun art project, why not make bubble prints? It's very easy, and campers love the messy fun. Be sure that you have a place to hang the artwork to dry, and place a plastic drop cloth or trays under your work area. This activity should only be done with young children if lots of supervision is possible. There is a chance they could accidentally drink the bubble mix.

Materials

- Shallow bowls
- Bubble mix, enough to fill each bowl about three-quarters full
- Food coloring or nontoxic liquid watercolor paint
- Drinking straws
- Copy paper
- Safety gear such as smocks and goggles, if desired
- Clothesline and clothespins

Process

1. Add bubble mix to each bowl. Add food coloring or liquid watercolor paint. Make sure that the bubble mix is fairly dark in color, and test it on white paper. You may need more colorant than you expect. Mix well.
2. Gently blow air into the bubble mix using the drinking straw. Try to create a nice rounded mass of bubbles over the liquid rising 1–2 inches above the bowl, without having the bubbles overflow the sides of the bowl.
3. Before the bubbles collapse, gently lay the paper over the bubbles. Try not to push it into the liquid. Lift the paper. There should be a colored print of where the bubbles popped against the paper. If it is faint, you may need to add more colorant.
4. Repeat until the paper is filled, switching colors as desired.
5. Clip the artwork to the clothesline, and allow it to dry completely.

Bubble Glow Sword

COST: $

MAKE TIME: 10 minutes

This was a last-minute idea one year for a science fiction–themed day at Maker Camp and ended up being a giant hit. The campers really wanted to have a sword fight, so I came up with an inexpensive way to do it. After making the glow sword, turn out the lights and use a long-exposure app on your phone for some epic pics.

Materials

- Store-bought giant bubble wands, the kind that come in their own long tube of bubble mix with a sturdy handle that twists in (Buy a party pack to keep the cost down.)
- Thin glow-in-the-dark glow sticks, the kind used for bracelets or necklaces (Buy a party pack. You'll want roughly three to five per camper.)
- Silver, gold, and black duct tape

Process

1. Have campers select their bubble and glow stick color.
2. Activate the glow sticks. Push them into the bubble solution. (You may need to pour a little out first.)
3. Use the duct tape to cover the handle, making it look like a light saber hilt.
4. Turn off the lights, and have epic slow-motion light saber battles. Then go outside and make bubbles (Figure 7.34).

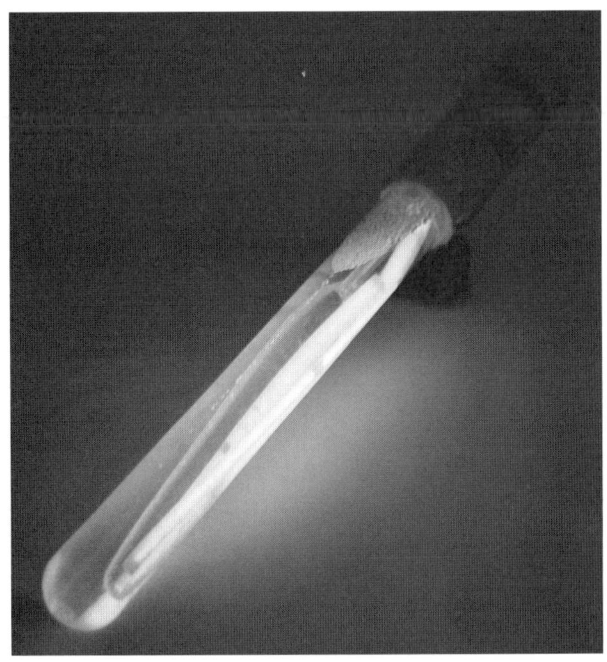

FIGURE 7.34

PROJECT 5 Smashed Flower Scavenger Hunt

COST: $

MAKE TIME: 10–15 minutes

This may be one of the crazier projects I've done, which is saying something. However, it is such a blast, it's absolutely worth doing. Not only does is get campers outside, but it also gets them really looking at nature. When we're all done, everyone has a unique collage to take home.

For the scavenger hunt portion, I usually make the list simple, such as a list of colors or differently shaped leaves. If you are very familiar with the area you're scavenging or want to teach specific wildflowers, trees, or plants, this is a great opportunity. Give campers paper lunch bags or something similar for collecting items. I usually set a time limit for the search and encourage campers to work with a friend. Then I plan a little time for the campers to share what they've found.

When selecting items, it's best to get the freshest plants possible. Flat items work best for this, but a little dimension is okay. Twigs and dry grasses won't yield great results. We'll be working on roughly 9- × 12-inch spaces, so campers should keep this in mind as they collect items as well. I always bring in a few extra bouquets of flowers and bundles of herbs from my garden (or the supermarket) as backups.

We'll be using rubber mallets, so before doing this activity, you'll want to make sure that you review proper hammer safety. Remind campers to always check around them and behind them before starting so that no one is accidently hit. Show campers how to grip the hammer tightly at the end of the handle with their thumb placed for stabilization. For this activity, they'll want to make many fairly light taps over the surface, so they may want to practice that before using their precious scavenger hunt finds.

Materials

- Scavenged flowers and leaves
- Heavyweight watercolor paper, cold pressed, 9 × 12 inches
- Pieces of plywood, roughly the size of the paper
- Rubber mallets
- Invisible or masking tape
- Binder clips or paperclips
- Scissors
- A soft paintbrush or toothbrush

Process

1. Decide on the size of the collage you'd like to make. You can either sandwich items between two pieces of paper or fold the paper in half and place items inside.
2. Lay out the items on your paper. Use tape to hold larger items in place as needed.
3. Use a clip to secure the edges of the paper. Place the paper on the plywood.
4. Using the mallet, begin hammering over the paper. It helps to start in the center and move outward in a circular motion. Be careful not to hit too hard because that can tear the paper.
5. Check the results. If needed, reclip and hammer some more. Different materials will need different amounts of force and repetition.

6. Once the camper is satisfied with the results, gently remove all the plant items. It can help to put items in a sunny spot to dry them or use a blow dryer set on low. Dried items are easier to remove. A soft paintbrush or toothbrush will help to remove dried items (Figure 7.35).

7. Once the print is dry, decorate it as desired with markers, pens, or colored pencils. These prints are especially nice when cut up to use as bookmarks.

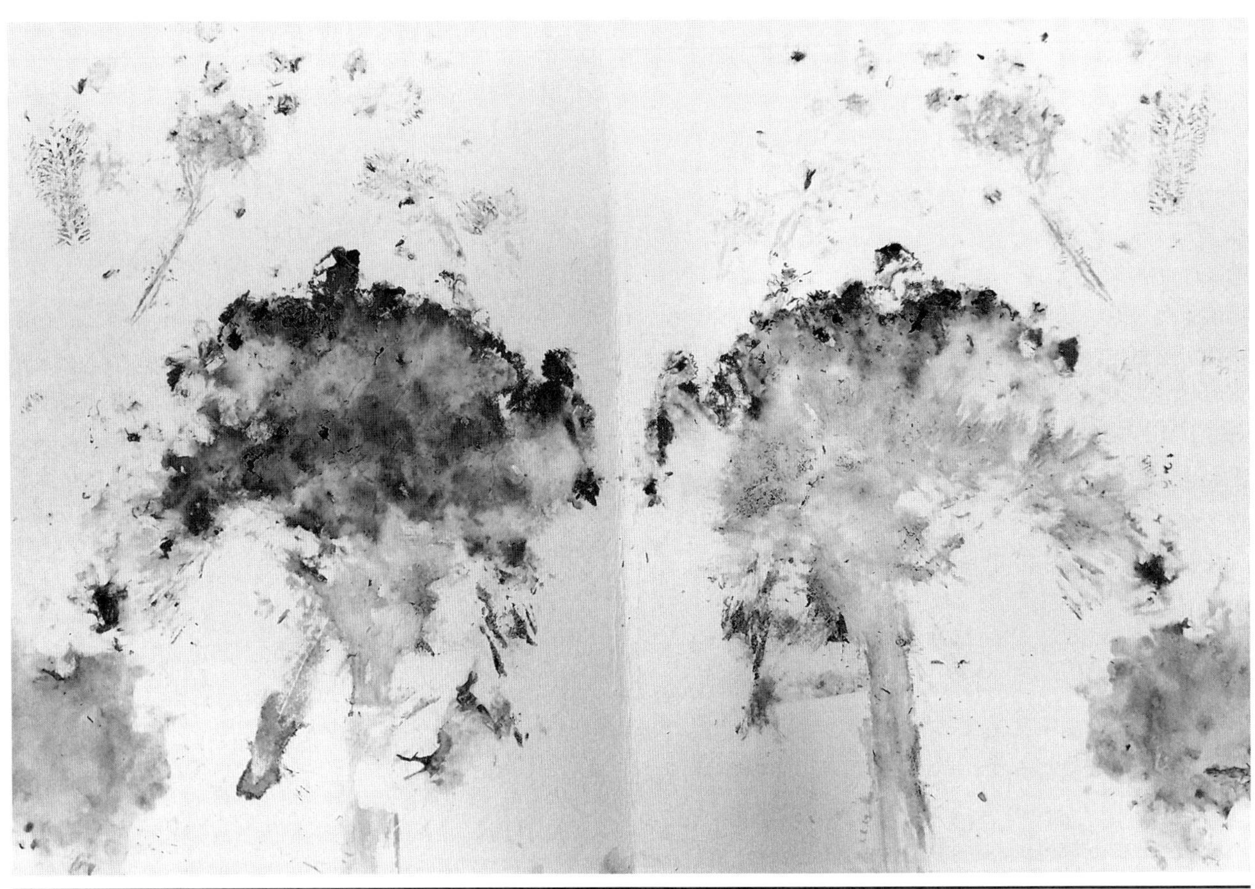

FIGURE 7.35

PROJECT 6
Seed Bombs

COST: $–$$

MAKE TIME: 15–30 minutes

Much like the "LED throwie" in the maker community, the seed bomb has a similarly rebellious history among the gardening set. Modern seed bombs are actually based on an ancient Japanese practice called *tsuchi dango*, meaning "Earth dumpling." The practice was reintroduced to the world in 1938 by Masanobu Fukuoka, a champion for sustainable agriculture. He used seed bombs as part of his natural farming method, bombing all kinds of areas with his creations and allowing vegetables to grow wild among the weeds, believing that Mother Nature and natural selection would yield the best plants without needing chemicals and lots of human intervention. Since then, would-be naturalists have used seed bombs as ways to spread biodiversity, add plant life to harsh or challenging habitats, help stop erosion, and bring natural beauty to urban landscapes.

When making seed bombs, you will want to be conscious of the materials you use. Spend the time and money to get quality native seeds meant for your area. Buy blends that don't include filler materials, if possible. You don't need much seed for each bomb, just a pinch. Use good screened compost, quality seed starting soil mix, and good topsoil rather than inexpensive blends with added fertilizers. Clay ideally should be dry clay meant for making seed bombs. Red clay is traditional. Select a natural, biodegradable clay with no additives. Many good distributors are available online, and your favorite local garden shop should be able to help. Wildflowers are the most traditional seeds for seed bombs these days, but herbs, clovers, groundcovers, and flowers to support local pollinator species are all excellent choices as well.

The ratio of clay to compost may vary based on your particular clay. The red clay I typically use calls for a ratio of five parts clay to one part compost, but others recommend a one-to-one ratio of clay and compost. Test in advance to make sure that your seed bombs hold their shape and harden completely without crumbling. I suggest starting with a three-to-one ratio and adding compost as needed. Often, campers add too much water to the mix, and additional soil will soak it up. Also keep in mind that red clay can stain.

Materials

- Native seed blend with plants meant for your local area
- Natural dry red clay
- Screened compost or quality seed starting soil mix without added fertilizer
- Reusable bowls and plates
- Spoons or Popsicle sticks
- Water
- Eye droppers or disposable pipettes
- Nontoxic tissue paper (optional)

Process

1. Place clay and compost in a bowl. Use the ratio suggested by your clay manufacturer, if possible. Otherwise, start with a three-to-one ratio of clay to compost, or 3 spoonfuls of clay and 1 spoonful of soil. Use your spoon or Popsicle stick to mix them well.
2. Using the dropper, add water slowly, 5 drops at a time, until the mixture is wet enough to form into a ball. Usually, you'll need about 20 drops, but the amount of water can vary widely based on your soil and clay.

3. Add your seeds. A good pinch of a flower blend is enough. For larger seeds (such as sunflowers), add two or three seeds total. Avoid adding too many seeds because the plants will crowd each other out when they germinate.
4. Roll the mixture into a ball. You can make one large Ping-Pong–sized ball or break it into smaller balls if you are using small seeds.
5. If desired, gently press tissue paper onto the balls, covering them (Figure 7.36).

FIGURE 7.36

6. Place the balls on a plate, and allow them to dry overnight or longer, as needed.
7. When the seed balls are dry, scout out places that could use some new plants. Do not seed bomb anyone's property without permission!

TIP: Although not as environmentally friendly, if you are working with a large group, you can mix everything in a snack-sized sandwich bag rather than a bowl to avoid dropping or spilling materials and to make cleanup easier.

Take It Further

Another fun way to make seed bombs is to use egg shells. Shake your egg to break the yolk. Use a pin to make a small hole in the pointed top of the egg and a larger hole on the rounded bottom. Blow through the smaller hole to push out the egg white and yolk. Rinse the shell well, and let it dry completely. Cover most of the egg with tissue paper using a one-to-one water and white glue mixture. Using a small funnel, add compost or soil with seeds through the bottom of the egg. Cover the rest of the egg, and allow it to dry completely. To use, crush the shell gently in your hand and toss.

PROJECT 7 Sun Prints

COST: $

MAKE TIME: 30–45 minutes

Maker Camp is a great time to get outside and enjoy the sun, but did you know that you can use those light waves to make art? You can! This project is quick, easy, and cheap to do, but it also demonstrates some really interesting science.

Common sun prints are actually more properly called *cyanotype photography*. Developed in the 1800s, the paper is coated with blue light-sensitive chemicals, which react to light waves and particles when exposed to light. Commonly known as *Berlin green*, the iron-based chemical compound physically changes when exposed to ultraviolet (UV) light, becoming a different chemical dye known as *Prussian blue*.

Prussian blue is very similar to Berlin green, but with one significant difference: it's not water soluble. So, if you cover areas of the Berlin green–coated paper with objects that block light, the area under those objects doesn't undergo the chemical change and remains Berlin green. The areas exposed to light do undergo the reaction and become Prussian blue. When you rinse the paper in water, the Berlin green is removed, leaving behind only the Prussian blue. This results in the characteristic white-on-blue silhouette of a sun print.

The objects you use can be just about anything as long as they block light. Because it's camp, it's especially fun to use this opportunity to explore nature. For example, I love to use field guides or internet resources to give campers a visual guide to different types of leaves common to the area. You can do the same for grasses, flowers, or other items. Indoor items such as keys, string, lace, puzzle pieces, game pieces, and other similar objects work well.

If you want to extend the science exploration for this activity, consider colored light bulbs, black lights, heat lamps, fluorescent lights, LEDs, and other types of light sources to which to expose the paper. If working outdoors, you can use colored acetate sheets over the sun prints to experiment with different wavelengths of light.

Materials

- Sun-sensitive paper
- Acrylic sheets, plastic wrap, or another clear material to hold objects in place
- Objects to use for print
- Water
- Lemon juice (optional)
- Plastic tub (optional)

Process

1. Collect items for the print.
2. Find a well-lit area with direct sunlight. Remove the sun-sensitive paper from the foil package, and place in the light. Working quickly, place the items you have collected on the paper.
3. Cover the items with a piece of clear acrylic or plastic wrap, using small rocks, if needed, to hold the paper and items in place (Figure 7.37).

FIGURE 7.37

FIGURE 7.38

4. Expose the paper to sunlight as directed by the instructions for your paper. The cloudier the day, the longer the paper will need for exposure. Usually, on a bright day, you only need about 5 minutes to expose the paper.
5. Rinse the paper with water. Again, check the instructions for your specific paper. With a large group, it can be easiest to fill several plastic bins with water for rinsing. Adding a few drops of lemon juice to the water can make the prints more intense.
6. Once rinsed, hang the paper to dry outside (Figure 7.38).

Take It Further

If you don't have access to sun-sensitive paper, you can place items on common construction paper instead. You will need to expose the paper for several hours, and the results are not as dramatic, of course, but it is inexpensive and doesn't require a rinse step. Darker colors work best.

You can also place squares of paper in plastic zip-top bags and then coat the bags with different levels of sunscreen. After exposing their prints, campers can compare results. Do different sun protection factor (SPF) levels really make a difference?

PROJECT 8
Pinhole Cameras

COST: $

MAKE TIME: 15–20 minutes

With all the emphasis on technology these days, it can be so important to remember that science and engineering weren't invented in the modern age. Each year for Maker Camp I like to throw in opportunities to explore the roots of our modern gadgets, such as with projects such as this pinhole camera.

Developed by Leonardo DaVinci, the *camera obscura* (Latin for "dark room") literally were dark rooms with a small hole on one side. The hole allowed a narrow beam of light to enter. The beam carried the image of outside objects onto the wall opposite the hole.

DaVinci was interested in this idea because the common belief at the time was that the eye worked by sending some kind of beam of energy outward, which bounced off of objects and returned to the eye. The camera obscura disproved that idea. It showed that light entered the eye and helped us to see.

Soon the camera obscura became the pinhole camera. Granted, the early pictures were upside-down, reversed images. As the light is reflected from the object through the small hole, it is distorted and flipped. But with the addition of a few mirrors and some very complex chemistry, Joseph Nicephore Niepce was able to take the first photographs in 1827. It was a revolution! So, in this project we celebrate this long development of a vital technology, which still impacts our lives to this day, with the humble pinhole camera.

Materials

- Paper towel rolls
- Wax paper
- Aluminum foil or black construction paper
- Duct tape
- Pins
- Scissors
- Ruler or tape measure
- Pencil or pen

Process

1. Use the ruler to measure 2 inches from the top of the paper towel roll, and make a mark. Use scissors to cut the top 2 inches off the tube.
2. Cover the new end with wax paper, taking care not to wrinkle it. Secure with duct tape (Figure 7.39).

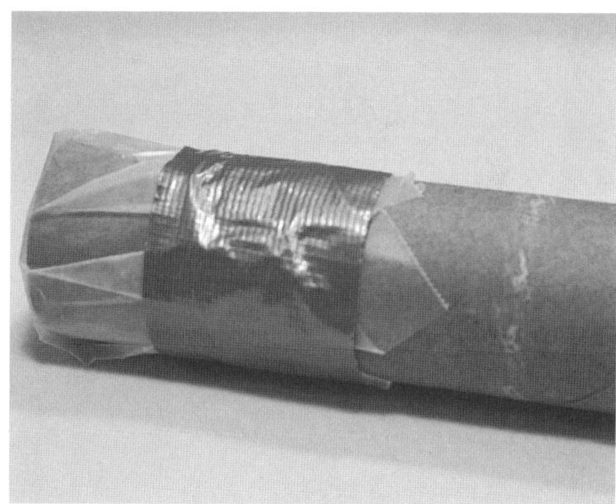

FIGURE 7.39

3. Place the cut-off piece on top of the wax paper end, and secure it with duct tape (Figure 7.40).

FIGURE 7.40

4. Cover the new end of the tube with aluminum foil, taking care not to wrinkle it. Secure the foil with duct tape.

5. Using the pin, make a small hole in the center of the foil (Figure 7.41).

FIGURE 7.41

6. If desired, cover the sides of the tube with foil to block additional light.

7. Either go outside or point the viewer out a well-lit window. Look to see the image reflected on the wax paper.

PROJECT 9
Live Model Stop Motion

COST: $$$

MAKE TIME: 60–90 minutes

Stop motion is one of the oldest animation techniques out there. It was first used for public showing in *Humpty Dumpty Circus*, created by Albert Smith and Stuart Blackton in 1899. Since then, it's been used in movies from the original *King Kong* through modern classics such as *The Nightmare Before Christmas*. Now there are many examples of stop-motion animation in commercials, cartoons, and other media.

The basic process involves taking a photograph of your objects or characters, moving them slightly, and taking another photograph. This is repeated over and over again, with just minor changes between each photo. When you play back the images consecutively, the objects or characters appear to move on their own.

Most commonly, artists create puppets with metal armatures that can be posed as needed. In Maker Camp, campers often enjoy using drawings, modeling clay, small toys, and Lego minifigures to create their stop-motion animation. Just about anything can be used as characters, including the campers themselves!

For Maker Camp, I like to use the campers as characters because the method lends itself easily to heading outdoors and using natural backdrops for animation. Rather than working on the small scale often associated with stop motion, campers have the opportunity to make their animation bigger and better. This can be especially fun if they create cardboard cities to trample like Godzilla or shift perspective by using the green grass as a backdrop and laying on the ground to create their scenes.

To create your animation, you'll need some kind of digital app to combine images into a movie. Free apps such as Stop Motion Studio on the iPad and Stop Motion Maker for Android allow campers to use their tablets and cell phones to create their movies. Both apps include an "onion skin" feature that shows a faint picture of the previous image so that you can more easily line up your next shot.

Of course, you can use any camera or smartphone to take the pictures and then import them onto a computer for editing. Software such as iMovie, Adobe Premiere, Windows Movie Maker, and Windows Photos all allow users to import photos and create a movie from them. They also allow campers to edit photos, add opening or closing credits, and add background music. Of course, this option doesn't include the onion skinning ability, which can be a challenge for some campers.

Then there are software options designed for educational use. My favorite is HUE Animation Studio ($19.99; educational discounts available). The software is easy to use and includes onion skinning and editing features as well as the ability to export to a variety of formats. You can use either the HUE HD digital camera or the webcam on a laptop to create video. For some campers, setting up a laptop and being able to photograph, edit, and share all in one spot is desirable.

Before starting, I suggest that campers take the time to brainstorm and complete a storyboard. A storyboard includes small pictures with worded descriptions to show the overall story in a step-by-step manner. Taking the time to plan a bit often yields better results and reduces creative conflicts among campers.

You'll also want to discuss the number of pictures campers will need to take to make their movies a reality. Stop motion uses the abbreviation *fps* (frames per second) to describe the speed at which the pictures are shown in the movie. At 10 fps, a fairly slow stop-motion rate, you'll need 10 pictures of every second of animation, or 600 for a full minute. This makes it even more important to plan ahead and test your work early and often before committing to a large project.

When the movies are finished, plan some time to make popcorn and view your creations together, if at all possible. Before sharing the creations online in any way, make sure that you have the permission of every parent, especially if your campers are 12 years of age or younger. I usually use Google Drive or DropBox to create a directory of movies for parents to enjoy.

Materials

- A tablet, smartphone, or camera
- A device to edit the animation
- An app or software to capture or create the stop-motion animation
- Props as desired

Process

1. Form a small group of two to four campers. Use a storyboard to plan the animation.
2. Assign characters. Make props as needed.
3. Photograph the images for the animation using the selected methods (described above). Combine the images in your software. Edit and add music as desired.
4. Share the movies with campers.

PROJECT 10
Making an Outdoor Game

It's too easy to stay inside at Maker Camp, delving into computers, crafting, and more. However, an essential part of summer for kids is getting outside and running around. There is no reason why you can't make game play a Maker Camp experience! Here are some ideas to try.

Make an Obstacle Course

COST: Free–$

MAKE TIME: 15 minutes to create courses, more to try them all

In this activity, two teams use similar items to create simple obstacle courses for one another. Make sure that you set some basic safety ground rules and give everyone a time limit before beginning, or things can get quickly out of control. You'll also want to discuss what kinds of obstacles are appropriate for various age groups so that no one creates an impossible course, accidental of otherwise.

The materials listed are suggestions. Just about any playground equipment can be used to create a course. The goal is to look at items creatively and try to use them in interesting new ways. You'll need a fair amount of space for this activity, but it's worth it. If you have a large group of campers, have those waiting to build watch and brainstorm. Then they can be the cheering squad as fellow campers go through the courses.

Materials

- Traffic cones
- Hula hoops
- Jump ropes
- Cinder blocks
- Balance boards or wooden planks
- Pool noodles and tent stakes for creating arches
- Sawhorses
- Tires
- Tennis balls to throw at targets
- Disposable water bottles and water or sand to act as targets
- Assorted buckets
- Burlap sacks

Process

1. Share the available materials with all campers, and review safety rules.
2. Divide into groups of three to four campers.
3. Give groups 5–10 minutes to brainstorm or plan.
4. Give groups 10–15 minutes to build their course.
5. Have each group compete on another group's obstacle course (Figure 7.42).

FIGURE 7.42

Water Gun Battles

COST: Free–$

MAKE TIME: 30 minutes to prepare the course; time to play varies.

There is nothing better on a hot day than to get outside and have an epic water gun battle. Especially if you happen to be working in a building without air conditioning, this simple activity can provide a lot of relief. And best of all, it isn't expensive or difficult to do.

Because some campers have access to large supersoakers and the like, whereas others don't, I always provide the water guns. Often, I'll pick up inexpensive versions at the store, spray them with white paint made for plastic, and let the campers decorate them with permanent markers the day before. That way, everyone has the same water gun to work with, and they have something fun to bring home. The campers especially love to have team members sign their water guns as keepsakes.

While a simple water gun battle is fine, I've found that my creative campers are always ready to take it up a notch, so I've started taking my cue from video games. I pick a wooded area or garden with some natural obstacles and cover. Buckets of water for refills will be hidden in the area. (A team automatically loses the battle if they dump the water purposefully.) Sometimes I hide better water blasters on the course as well. Every camper wears safety goggles, and we discuss safety before we begin.

To add a maker component, I'll give campers some basic supplies to make a simple lockbox or trap. Into the trap, I place Popsicle sticks, each a different color for each team of two to three campers. These will be hidden around the course as well. The object of the battle is to collect all the colored sticks without damaging the holder and return to home base first. Of course, because everyone gets to run around and get wet, really everyone wins!

Materials

- Safety goggles
- Simple water guns, one for each camper
- Colored Popsicle sticks
- Cardboard boxes or scrap cardboard
- Paper and foam cups
- Paper and foam plates
- Paper and plastic bags
- Toilet paper and paper towel tubes
- Rubber bands
- String or yarn
- Drinking straws
- Popsicle and craft sticks
- Bamboo skewers or wooden dowels
- Assorted tape: invisible, duct, masking
- Assorted glues: hot glue, white glue
- Paint and markers
- Buckets for water

Process

1. Divide campers in to groups of two to three.
2. Give campers 15–20 minutes to build a contraption to hold their team's colored Popsicle sticks while making them difficult to get. This may mean devising a lock, creating a hidden compartment, camouflaging the box, or even creating a method to place the box in a tree. The stick holder must be able to be opened without being damaged so that teams have a fair shot at getting the sticks. You may want to do this the day before if paint or glue needs the chance to dry.

3. Hide buckets of water and traps on the course. Set a home base.
4. Have everyone put on safety gear. Remind campers that they should not aim for the head but rather the chest. No one is to dump the buckets of water intentionally. Once a team discovers its trap's location, it is allowed to defend the location, but only by shooting water. Teams may not block an opposing team in any other way. At no time may campers physically touch one another while on the course.
5. Start all campers from home base at the same time. The first team to return with every color of Popsicle stick wins the battle. All members of the team must cross the line for it to be considered a win.

Pool Noodle Hockey

COST: $

MAKE TIME: 15–30 minutes to prepare; time to play varies.

This game is a great way to use up some energy at the end of the day. Campers will build their own hockey sticks and then battle to get as many whiffle balls across the line as they can.

Materials

- Pool noodles, foam pipe insulation, and similar materials
- Scrap cardboard
- Duct tape
- Whiffle balls, 40–50
- A clothesline or long piece of rope
- Buckets or plastic bins

Process

1. Give campers 10–15 minutes to create their hockey sticks. Sticks must be soft enough that they cannot injure anyone who is accidentally hit.
2. Use the clothesline to create a large circle as the court.
3. Scatter the whiffle balls across the court.
4. Place a bucket or bin as home base for each team. Play with at least two teams, no more than four depending on the size of your circle and number of people on each team.
5. All campers must start on the outside of the circle. When the "Go" is given, campers run into the circle and use their hockey sticks to push the balls into their team's bin. They may not touch the balls with anything except their hockey stick, and the balls may not leave the ground.
6. Once all the balls are out of the circle, the balls in each team's bucket are counted. The team with the most balls wins.

Blacktop Games

COST: Free

MAKE TIME: 15–30 minutes each

You don't need fancy supplies to create a great game. With some chalk on a blacktop, campers can re-create and adapt many classic playground games. Although I present the basic rules, there is no reason campers cannot change the rules as long as everyone playing agrees. These games are inexpensive, easy to set up, and need minimal materials.

Four Square

Materials

- Chalk
- A kickball, volleyball, or similar ball

Process

1. Draw a large square that is subdivided into four smaller squares. Number the squares 1–4, going clockwise.
2. Each player stands in one of the four squares.
3. To start the game, the player in square 1 serves the ball by bouncing it in his or her square once and then hitting it toward one of the other squares.
4. The receiving player then hits the ball to any other player in one of the other squares. The ball may not bounce more than once in the receiving player's square.
5. If a player hits the ball so that it doesn't enter another player's square, he or she is "out." If the player fails to hit the ball before the second bounce after it has landed in his or her square, he or she is "out."
6. When a player is out, he or she must leave the square. The player in the next square moves up, as do the other players. A new player takes the number 4 position.
7. The object of the game is to move up to and hold the server's position for as long as possible.

Hopscotch

Materials

- Chalk
- A small rock

Process

1. Lay out your court. A number of layouts have been used, but generally courts feature a mix of single and double blocks. For example, a typical layout may be three blocks stacked on top of one another vertically, labeled 1, 2, and 3; then two blocks placed side by side, labeled 4 and 5; then two more vertical blocks, labeled 6 and 7; then two side-by-side blocks, labeled 8 and 9; and finally, one wide block labeled "Home." Campers should be encouraged to create new layouts once they have the basic idea.
2. The player tosses a marker or small rock onto the court. The squares are numbered in the sequence in which they are to be hopped. The rock must land within the correct square each time.
3. The player then hops through the court, skipping the box where the stone landed. For single boxes, the player must hop on one foot. For side-by-side double boxes, both feet much touch the ground at the same time with left foot landing in the left square and the right foot landing in the right square.

4. When a player reaches the Home square, he or she must turn around and return through the court. The player must stop in the square before the marker. The player must reach down to retrieve the marker and continue the course as started, without touching a line.
    a. If, while hopping through the court in either direction, a player steps on a line, misses a square, or loses balance, his or her turn ends.
    b. If a player successfully completes the sequence, the player continues the turn by tossing the marker into square number 2 and repeating the pattern.
5. The first player to complete one course for every numbered square on the court wins the game.

Freestyle

Give campers the bin of chalk, and invite them to create mazes and artwork on the blacktop. Try drawing a map of your location.

PROJECT 11
Maker Rocks

COST: Free–$

MAKE TIME: 15–30 minutes to make rocks; time to hide and find rocks varies.

"Kindness Rocks" have taken the country by storm. The idea is very simple: paint rocks and leave them for others to find. It's become a type of scavenger hunt focused on spreading joy, hope, encouragement, and beauty. The first Kindness Rocks were left by Megan Murphy on the beaches of Cape Cod and featured inspirational affirmation. She has now formed the Kindness Rocks Project. Their motto is, "One message at just the right moment can change someone's entire day, outlook, life." Groups have formed all over the country, with people taking pictures of their creations and sharing stories when they are found. You can learn more at www.thekindnessrocksproject.com/.

So why not put the same idea into motion for your young campers? Using the same ideas, encourage them to share what they love to make. Have them share their favorite math or science concept. Paint a robot, computer, or bridge of Bucky ball. Or why not feature a quote from a favorite maker? Painting "Maker Rocks" is not only a ton of fun, but it is also an opportunity for your campers to identify with what makes them makers and to celebrate it by sharing that truth with others.

It's also a fantastic chance to get outside! Usually, we paint our rocks first thing in the morning. By afternoon, they are dry, and we can hide them before leaving camp. The next morning, I set the campers loose to find rocks, snap pictures, and then rehide the rocks. We'll repeat that cycle as often as they like, often adding additional rocks as we go. At the end of camp, everyone takes home a rock that has inspired them as a keepsake, and everyone gets access to the directory of photos to enjoy.

Just remember that when you hide rocks, you need to make sure that you have permission. Public spaces are fine, but campers should be reminded not to trespass on private property. Also, while it's great to encourage campers to be creative with where they hide their rocks, you want to remind them to keep the locations safe and accessible to every camper. (This is especially important if you have campers with disabilities.)

You may also want to include contact information for your organization or camp on the rocks so that if any are left behind for others—and I encourage to leave some behind—they have a way to learn more about Maker Camp. You can also include the Maker Camp website (www.makercamp.com) in hopes of inspiring others.

Materials

- Smooth rocks, several inches in diameter and fairly flat if possible
- Acrylic paints, including glow-in-the dark, metallic, and glitter paints
- Paintbrushes
- Hair dryer (optional)
- Permanent markers
- Clear acrylic spray sealant
- Paper plates
- Smartphones, tablets, or cameras to take pictures

Process

1. Place your rock on a paper plate. Use an additional plate as a palette for your colors.
2. Paint your rock. Use thin layers for quick drying. If needed, use a blow dryer set to cool to aid in the drying process.
3. If you don't love using acrylic paints, try permanent markers instead. You don't get as much coverage as you do with paint, but the rocks are ready to go almost immediately.
4. When the rocks are finished and dry, spray on two to three light coats of clear sealant, letting the rocks dry between coats. The artwork on the rocks will not survive the elements otherwise.
5. Hide your rocks outside, documenting where they are with a photo. It can be fun to make up riddles about where they are hidden.
6. Share the photos with the group, and set them loose to find the rocks and hide them somewhere new, taking a picture each time the rock is moved.
7. Repeat as often as you like (Figures 7.43 and 7.44).

FIGURE 7.43

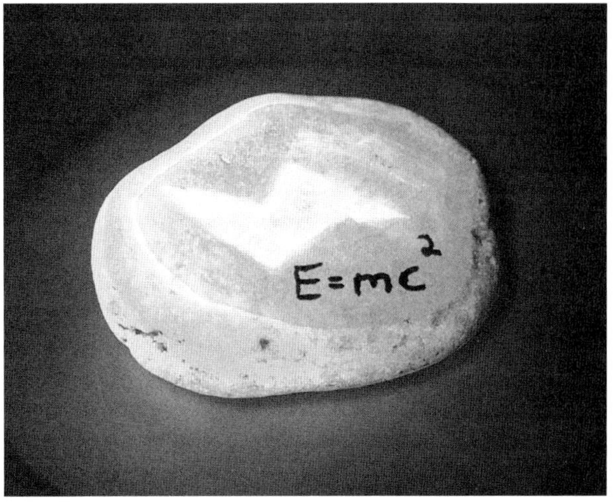

FIGURE 7.44

Take It Further

Get creative with painting your rocks by introducing one of these fun techniques:

- Mask areas of the rocks with small stickers, hot glue, masking tape, or rubber bands. Paint over the masks. Remove the masks before the paint dries.
- Place the rocks in a salad spinner. Add a few drops of paint to each. Spin in the spinner to spread the paint. Add more paint as desired (Figure 7.45).

FIGURE 7.45

- Heat the rock with a craft heat gun. Wearing heat-safe gloves, draw on the rock with wax crayons. Heat again to make the colors melt and blend. Add layers until you are satisfied (Figure 7.46).
- Use a mixture of glue and water to adhere torn pieces of tissue paper, magazine pages, or old sewing patterns to your rock.

FIGURE 7.46

CHAPTER 8

Making Your Community

MAKER CAMP IS MORE THAN THE PROJECTS YOU complete with your campers. It's about learning and growing as a maker, building skills, solving problems, adding a bit of yourself to your projects, and overcoming challenges. I remind my young campers all the time to focus on "process over product." At the end of the day, how you make the thing is more important than what you made (or failed to make). Our goal is not to make stuff; it's to make makers.

One way you can keep campers focused on that process is to spend time telling the story behind their creations. Through photography, drawing, and journaling, you can highlight the brainstorming sessions, the prototyping, and the construction, as well as the moments of wonder and laughter. When you share those items, you not only celebrate your camper's creativity and resilience, but you also welcome the world into your Maker Camp journey and may inspire others to begin their own.

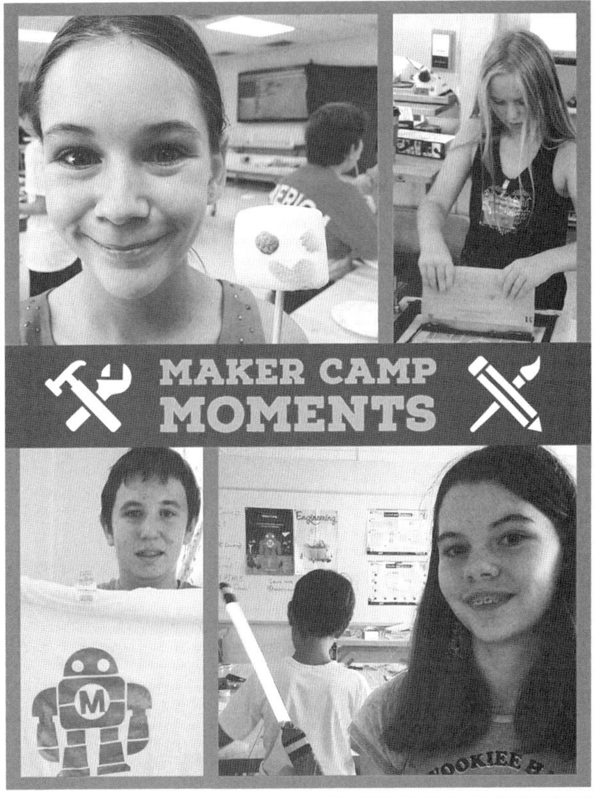

FIGURE 8.1

Your Maker Camp Scrapbook

Keeping a scrapbook is a timeless trend. Even before photography, people kept scrapbooks of artwork, letters, pressed flowers, and more. Over time, the scrapbooks themselves have been elevated into works of art, carefully crafted to tell a story and highlight important events. Of course, with the introduction of home computers, graphic arts software, and the internet, scrapbooks have gone digital.

As you go through your Maker Camp session, consider compiling a collection of great pictures into a shared "cloud" that can be accessed securely by campers. I tend to take pictures on my smartphone. Every night at home, they automatically upload to both DropBox and my Google Drive. This makes it fairly easy to create a folder that others can get to online while also controlling access. You can also encourage campers to upload their snapshots to share.

To make a paper scrapbook, simply put out 12- × 12-inch cardstock paper for campers to use as a base for their page. Provide stickers, interesting papers, die cutters, decorative scissors, markers, rubber stamps and ink, and other craft supplies.

PRO TIP: Ask families to donate supplies. Many folks have leftovers they are happy to pass along. You may want to take time to make your own rubber stamps or stencils as well.

Have your best pictures printed at a store that provides one-hour photos. Or if you have access to a photo printer such as the Canon Selphy, campers can select and print their favorites at camp. If you have instant cameras, such as the Fiji Instamax, you can collect prints all week.

When pages are complete, share them with others. Campers can decide if they want to keep the page or donate it to a scrapbook that grows from year to year. If campers take home their pages, be sure to snap a good picture of each or scan and copy each. They are great as examples for next year.

To make digital scrapbooks, all you need are computers and software. If your school or library has Microsoft Publisher, you can use it to create pages. Adobe Elements is also a great piece of software for making digital scrapbooks. You can also try free online services such as Canva or Adobe Spark to create pages. There are many sites that offer free scrapbooking elements (digital papers, characters, etc.) for download. Plan ahead, and have a shared folder with materials prepared.

Of course, if you are going digital and working with older campers, you may want to put a twist on your scrapbook by creating a zine instead. A zine is a small self-published magazine usually made by small groups to highlight something they love. Think of it like a hacked magazine. In addition to photos, campers can offer written reflections, original artwork, instructions to their own DIY projects, and more.

Digital Portfolios

If you want to take it to the next level, consider creating digital portfolios of each camper's work. A digital portfolio is simply an electronic collection of student work featuring a variety of media. In an academic setting, they are used to demonstrate growth, learning, and student achievement over time. Typically, they are shared with parents, other teachers, administrators, and even other students. They can be a fantastic way to assess student work in a classroom setting.

However, a digital portfolio can work well for informal settings such as camp, too. In this setting, the goal is again "process over product." Campers should be encouraged to document their work step by step telling the story of what they've made. They should share what worked and what didn't, reflecting on each. This builds the foundation of an important part of the maker mind-set: openly sharing your work with the community. It takes practice to learn how to communicate in this way, but it's a great skill to develop. Encourage campers to take their digital portfolios and use them to make full-fledged "how to" articles.

Maker Camp digital portfolios should be shared online so that family and friends can see their campers' work, but care needs to be taken to protect the identity of young campers. Make sure that whatever platform you select, you have permission from guardians. Take the time to review with campers what information is safe to share and what isn't. Common Sense Media offers a variety of curricula and resources on digital literacy that you may want to consider.

An easy way to design a digital portfolio is to use Google Sites (https://sites.google.com/). You can restrict access fairly easily, especially if you are using a school's Google Education server. Google Sites are free, easy to design, and easy to share. They have a variety of built-in templates that make it easy to create pages. From a design standpoint, Google Sites works in a similar manner to Google Docs or Slides, so the format is familiar to many campers.

Another great free platform for digital portfolios is SeeSaw (https://web.seesaw.me/), available as an app and online. SeeSaw makes it very easy for students to share work with the ability to take photos, write notes, add captions, link to online sites and files, and more, all in one place. It has some nice social media-style functions, too, letting campers add stickers to photos and "Like" one another's posts. You have complete control over what gets shared and who sees it, so it makes security easy. It also makes it easy to organize the work of multiple campers in one place (Figure 8.2).

Make Media also has a platform for sharing projects on MakerShare (https://makershare.com/). The big benefit to MakerShare is that it is ready made specifically to share maker projects. The templates are already there, and all you have to do is fill in each section. It's a great way for older makers to really organize their content.

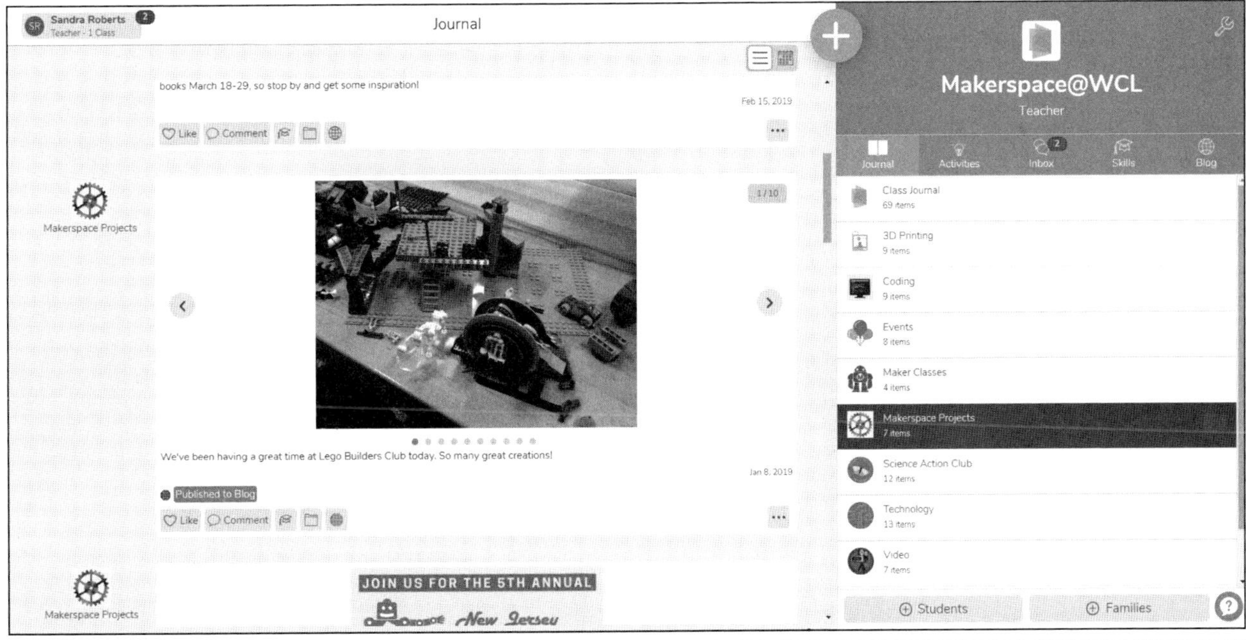

FIGURE 8.2

It's also a place the make editors tend to check for interesting projects, which can lead to your campers' work being shared on an even larger platform.

Regardless of the format you choose, be sure to put the work of keeping the digital portfolio into the hands of your campers. Once you've shown them the basics, let them decide how much and what to share. Some campers will love the process; other may not. That's okay. Camp is not school.

You may also want to consider having one portfolio for camp as a whole. For example, in my makerspace, I keep two iPads with SeeSaw installed available to anyone who comes in. They document their work if they want, and selections get shared out to the web.

If you want to learn more about digital portfolios, KQED Teach has paired up with MakerEd to offer free online courses about using them (https://teach.kqed.org/course/digital-portfolios-with-maker-ed).

Sharing Your Maker Camp

Sharing pictures and ideas from Maker Camp is one of my biggest joys of summer. It's also a great way to build your maker community. Social media provides many options for both (Figure 8.3).

Many of your parents are likely on Facebook, so why not create a free Facebook group for your camp. In addition to posting photos and videos of the kids creating, you can post additional projects to do at home, keep up on scheduling and administrative details, answer questions, and encourage campers to stay connected even after camp is over for the summer. Forming a Facebook group is probably the best way to build your local community of makers.

Instagram is, of course, the perfect place to share your pictures and videos. It's also where most of your 13+ campers are probably spending their time. Ask them to follow your Maker Camp Instagram account. Encourage them to share what they are making through the platform. Put the selfie culture to good use

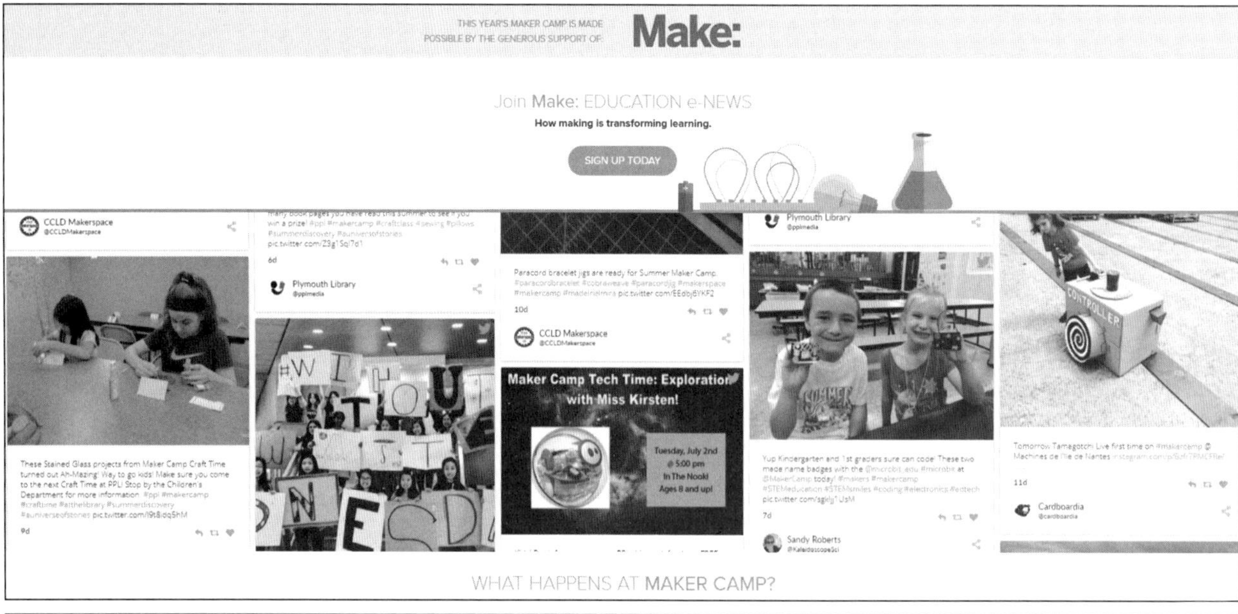

FIGURE 8.3

by having them capture a "maker moment" every day at camp. Using Instragram can be a way to empower your campers and push their perspective away from a world that can prize a consumer-based, glossy-surfaced fiction into one that shows a confident reality focused on making the world a better place.

And then there is Twitter. I love Twitter. Yes, it can be a bit overwhelming, especially at first, but there are so many amazing makers, educators, organizations, and companies out there to provide support and inspiration. Not only will they be genuinely excited to see what your campers are up to, but you'll also find people who are willing to offer advice and information while they cheer you on. You can start by following me at @KaleidoscopeSci. I'll always be there to help you out and give you likes or retweets.

I look forward to seeing what you and your campers make this summer and for many summers after (Figure 8.4).

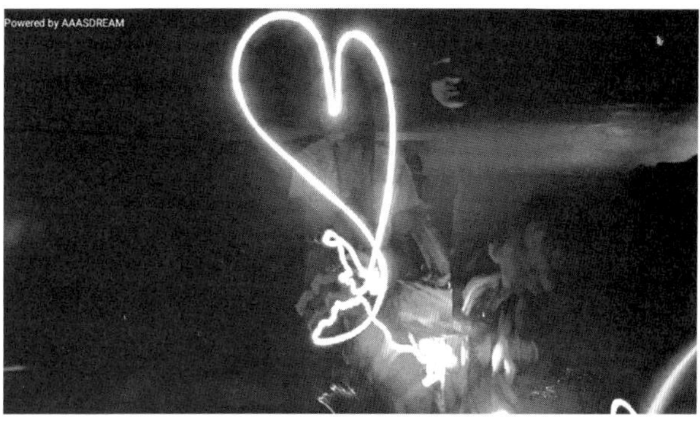

FIGURE 8.4

Index

Numbers

3D printing
  backpack tags, 62–65
  bubble wands, 234–240
  chocolate molds, 182–186
  cookie cutters, 187–202
  CPX holders, 42
  making leather stamps, 49
8-Step Drummer, 170
16-Step Sequencer, 171

A

ABS, 182
Acurite, 220
Adafruit Industries, 42, 128
Adobe Elements, 262
Adobe Spark, 262
affiliates, 3
allergy issues, 173
amperage, 77
anemometer, 218–219
Arduino, 221
Audacity, 167
Augmented Reality (AR), 62
Austin, John, 145

B

backer rod, 106
backpack tags, 62–65
Banana Piano, 152, 157
barometer, 217
Bengala dyeing, 32
  See also dirt shirts
Berlin green, 247
Big Book of Makerspace Projects, The (Graves), 78
bit depth, 167
Blackton, Stuart, 251
blacktop games, 255
bleached shirts, 25–28
bracelets
  stamped leather bracelet, 48–49
  ultraviolet color-changing sun bracelet, 38–39
Branwyn, Gareth, 221
British Broadcasting Company (BBC), 84
brown sugar chocolate molds, 180–181
bubble projects
  3D-printed bubble wand, 234–240
  bubble blowers, 231
  bubble glow sword, 241–242
  bubble nets, 231–234
  bubble prints, 241
  bubble wands, 231
  chenille stem wands, 231
  general information, 229
  giant bubbles, 229–230
  making the bubble mix, 229
budgeting projects, 4–6
Burgess, Phillip, 128

C

CAD software, 62
  See also Tinkercad
cameras, 74
  camera obscura, 249
  DSLR cameras, 150–151
  Fuji Film Instax, 69, 71
  long-exposure camera apps, 145
  pinhole cameras, 249–250

camper "Guess Who?" board game, 69–73
campfire popcorn, 207–208
campfire projects
  building a LED campfire, 126–134
  coding a robot drum circle, 159–163
  color-changing flames, 122–123
  decorative campfire display, 126
  individual campfires with Circuit Playground Express (CPX), 127–134
  individual campfires with tea lights, 126–127
  LED fireflies, 135–139
  Luna moths, 140–144
  making music with Makey Makey, 157–158
  making musical instruments, 152–156
  making up a maker camp song, 164–165
  mixing music with the NeoTrellis, 166–172
  ninja stars, 145–149
  roasting spy marshmallows, 124–125
  spooky ghost photos, 150–151
Canon Selphy photo printer, 69, 74, 262
Canva, 262
cardboard box guitar, 155–156
cardboard swords, 59–61
casting and molding projects
  chocolate, 177–186
  creating resin action figures, 114–117
caulk saver, 106
Ceceri, Kathy, 144
Chibitronics, 78, 83, 145
chocolate projects
  3D-printed chocolate molds, 182–186
  brown sugar chocolate molds, 180–181
  creating silicone molds, 181–182
  general information, 177
  tempering, 177
  using commercial molds, 178
  using cookie cutters as molds, 178–179
cicadas, 224
Circuit Playground Express (CPX)
  individual campfires with, 127–134
  in light-up proximity friendship necklace project, 42–47
Circuit Playground Jack-o'-Lantern, 128
CircuitPython, 42, 166, 168, 169
circuits
  paper circuits, 78–79
  parallel circuits, 78
  series circuits, 77
citizen science, 223–224
coffee can drum, 152
color-changing flames, 122–123
commercial molds, 178
community
  digital portfolios, 262–264
  Maker Camp scrapbook, 261–262
  sharing your Maker Camp on social media, 264–265
Connors, Chris, 15
Cookie Caster, 188
cookie cutter molds, 178–179
cookie cutters, 3D printed, 187–202
cookie recipe, 201–202
cooking projects, 173
  3D-printed cookie cutters, 187–202
  campfire popcorn, 207–209
  edible treehouse, 205–206
  making chocolate, 177–186
  making marshmallows, 174–176
  solar ovens, 203–204
  sorbet toss game, 210–211
cosplay, 37, 55
  books on, 55
  cardboard swords, 59–61
  masks, 55–58
  T-shirt capes, 58–59
CPR training, 7
CPVC, 109, 110
CPX. *See* Circuit Playground Express (CPX)
current, 77
cyanotype photography, 247–248

D

Da Vinci, Leonardo, 249
Dance Off, Makey Makey, 91–100
dance pad, creating, 98–100
Davis, 220
digital cutters, making stencils with, 11–14
digital portfolios, 262–264
digital scrapbooks, 262
dirt shirts, 32–35
Do Ink app, 74
dust masks, 6

E
ear plugs, 6
eyewear, 6

F
Facebook, 264
fireflies, 135–139
Flipgrid, 76
four square, 256
fps (frames per second), 252
Free Music Archive, 76
Fried, Limor "Ladyada," 42
friendship bracelets, 38–39
Fukuoka, Masanobu, 245
fun, 7

G
games
  blacktop games, 255
  camper "Guess Who?" board game, 69–73
  creating resin action figures, 114–117
  four square, 256
  green-screened camp photos and videos, 74–76
  hopscotch, 256–257
  Lego labyrinths, 68
  light-up letter home, 77–83
  Makey Makey Dance Off, 91–100
  marshmallow poppers, 113
  Nerdy Derby, 118–120
  Nerf wars, 101–111
  obstacle course, 111, 253
  outdoor games, 253–257
  pool noodle hockey, 255
  PVC marshmallow shooter, 112
  Smack a S'more, 84–90
  sorbet toss game, 210–211
  water gun battles, 254–255
Gemma, 52, 54
Geo-Wiki Picture Pile, 224
GIMP, 201
  designing a basic cookie cutter in, 197–200
GLOBE Cloud, 223
gloves, 6
Google Sites, 263
Graves, Colleen, 78
Green Screen Live Video Recorder, 74
Green Screen Video for Android, 74
green-screened camp photos and videos, 74–76
"Guess Who?" board game, 69–73

H
hopscotch, 256–257
Hue Animation Studio, 251
Humpty Dumpty Circus, 251

I
Illustrator, 188, 192–196
iMovie, 74
iNaturalist, 223
infrared radiation (IR), 42
inks
  mixing, 18–19
  See also screen printing
Inkscape, 188, 201
  designing a basic cookie cutter in, 192–196
Instagram, 264–265
IR. *See* infrared radiation (IR)

J
JavaScript, 42, 84
Jiffy Pop. *See* popcorn
Journey North, 223

K
Kindness Rocks Project, 258
King Kong, 251
KQED Teach, 264

L
leather bracelet, 48–49
LED crafts, 40–41
  building a LED campfire, 126–134
  fireflies, 135–139
  light-up bugs, 144
  light-up letter home, 77–83
  light-up sunglasses, 51–54
  Luna moths, 140–144
  ninja stars, 145–149
Lego labyrinths, 68
light-emitting diodes. *See* LED crafts
light-up bugs, 144
  See also fireflies; Luna moths

light-up letter home, 77–83
light-up proximity friendship necklace, 42–47
light-up sunglasses, 50–54
LilyPad, 52, 54
live model stop motion, 251–252
Long Expo, 145
Loop of the Loom, 32
Low Shutter Cam, 145
Luna moths, 140–144

M

magazines for makers, 4
Maillard reaction, 124
Make Magazine, 4, 15
MakeCode, 84, 85–90, 221–222
  in individual campfires with Circuit Playground Express, 128–130
  in light-up proximity friendship necklace project, 42–47
Maker Camp
  affiliates, 3
  overview, 1–3
  scrapbook, 261–262
  sharing on social media, 264–265
maker camp song, 164–165
Maker Faire, 4, 32
Maker Media, Inc., 3
Maker Movement, 2, 9–10
maker rocks, 258–260
MakerEd, 264
MakerShare, 263–264
Makerspaces.com, 78
Makey Makey, 83
  coding a robot drum circle, 159–163
  Dance Off, 91–100
  making music with, 157–158
Makey robot logo, 10
  making a stencil of, 11–14
  See also stenciling
marshmallow projects
  making marshmallows, 174–176
  poppers, 113
  roasting spy marshmallows, 124–125
  shooter, 112
masks, 55–58
micro:bit projects
  Smack a S'more, 84–90
  See also MakeCode
Microsoft MakeCode. *See* MakeCode
Microsoft Publisher, 262
microwave popcorn, 208–209
MIDI audio board. *See* NeoTrellis
Mini Weapons of Mass Destruction (Austin), 145
molding and casting projects
  chocolate, 177–186
  creating resin action figures, 114–117
Mono, 167
Morphi, 62, 188
Mosquito Byte app, 224
Mr. Makey tag, 65
mucilage, 174
Murphy, Megan, 258
musical projects
  coding a robot drum circle, 159–163
  making music with Makey Makey, 157–158
  making musical instruments, 152–156
  making up a maker camp song, 164–165
  mixing music with the NeoTrellis, 166–172

N

National Phenology Network, 223
Nature's Notebook, 223–224
necklaces, 42–47
Nelson, Carter, 42
NeoTrellis, mixing music with, 166–172
Nerdy Derby, 118–120
Nerf blasters
  cosmetically enhancing, 101–105
  functionally enhancing, 105–110
Nerf targets, 111
Nerf wars, 101–111
NerfHaven, 108
nested loops, 45
Nightmare Before Christmas, The, 251
ninja stars, 145–149

O

obstacle course, 111, 253
  See also Nerf wars
ohms, 77
Online-Convert, 187–188

origami projects
  LED fireflies, 135–139
  Luna moths, 140–144
  ninja stars, 145–149
O-rings, 108
outdoor projects
  bubbles, 229–242
  citizen science, 223–224
  games, 253–257
  live model stop motion, 251–252
  maker rocks, 258–260
  pinhole cameras, 249–250
  PVC stomp rocket launcher, 225–228
  seed bombs, 245–246
  smashed flower scavenger hunt, 243–244
  sun prints, 247–248
  weather stations, 214–222

P
paper circuits, 78–79
Paper Inventions (Ceceri), 144
paper plate tambourine, 154
parallel circuits, 78
Parks, John, 170, 171
PCM, 167
permanent marker "tie dye," 22
photobooths, 75
photography projects
  camper "Guess Who?" board game, 69–73
  cyanotype photography, 247–248
  green-screened camp photos and videos, 74–76
  pinhole cameras, 249–250
  spooky ghost photos, 150–151
  See also cameras
Photoshop, 197–200
Photoshop Premiere, 74
pinhole cameras, 249–250
Pixabay, 76
PLA, 62, 182, 188
planning projects, 4–6
Plumber's Goop, 109
PodSafe Audio, 76
pool noodle hockey, 255
popcorn, 207–209
portfolios, digital, 262–264
positive, 114
precaulking rope, 106
prop-making resources, 104–105
Prussian blue, 247
pulse-code-modulation, 167
PVC, 109
  marshmallow shooter, 112
  stomp rocket launcher, 225–228

Q
Quindt, Svetlana, 104, 105

R
rain gauge, 214–215
rain sticks, 154–155
Raspberry Pi, 221
recipes
  cookies, 201–202
  marshmallows, 174–176
  popcorn, 207–209
  royal icing, 206
  See also cooking projects
Rembor, Kattni, 42
resin action figures, creating, 114–117
resistance, 77
resources
  on Circuit Playground Express (CPX), 42
  on cosplay, 55
  on digital portfolios, 264
  on the Maker Movement, 2–3
  on makerspace projects, 78
  on Makey Makey, 163
  on musical instruments, 156
  for NeoTrellis code and sound files, 166, 168, 171–172
  on origami, 144
  on prop-making, 55, 104–105
  on safety, 7
  for songs, 164
  on Steampunk, 55, 104
roasting spy marshmallows, 124–125
robot drum circle, 159–163
rocket launcher, 225–228
royal icing, 206

S

safety, 6–7
sample rate, 167
scavenger hunt, 243–244
Schwartz, Marc-Olivier, 221
scrapbook, 261–262
Scrappy Circuits, 83
Scratch, 91–98, 159–163
screen printing, 15–21
seed bombs, 245–246
SeeSaw, 76, 263
sequencer, 171
series circuits, 77
Shapeways, 62
Silhouette digital cutter, 57
Silhouette Studio, making a stencil with, 11–13
silicone molds, 181–182
silkscreens
  applying a vinyl stencil to, 17–18
  making, 16–17
  See also screen printing
SketchUp, 62
Smack a S'more game, 84–90
smashed flower scavenger hunt, 243–244
Smith, Albert, 251
social media, 264–265
soda ash, 32, 33
solar ovens, 203–204
sorbet toss game, 210–211
sound files, 167–168
sound mixer, 170
SparkBooth, 74
SparkFun, 83
spin-art machine, building, 29–30
spin-art T-shirts, 29–31
spooky ghost photos, 150–151
spray-painted shirts, 23–24
spring index, 107
staff, 5
stamped leather bracelet, 48–49
Stanford SOLAR System, 38
Steampunk resources, 55, 104
stefans, 105–107
stenciling
  bleached shirts, 25–28
  dirt shirts, 32–35
  with permanent markers, 22
  screen printing, 15–21
  spin-art T-shirts, 29–31
  spray-painted shirts, 23–24
stencils, making with a digital cutter, 11–14
stop motion, 251–252
Stop Motion Maker, 251
Stop Motion Studio, 251
sun prints, 247–248
sunglasses
  glow stick sunglasses hack, 50–51
  glow-in-the-dark painted sunglasses, 50
  light-up sunglasses, 51–54
SVG files, 187–188
swords, cardboard, 59–61

T

thermometer, 216
Thingiverse, 42, 65, 186
Thorsson, Shawn, 104, 105
tie dye, with permanent markers, 22
Tinkercad, 62–65, 182–186, 187
  3D-printed bubble wand, 234–240
  designing a cookie cutter in, 189–191, 195–196, 199–200, 201
T-shirts, 9–10
  bleached, 25–28
  capes, 58–59
  dyeing with dirt, 32–35
  making stencils with a digital cutter, 11–14
  permanent marker "tie dye," 22
  screen printing, 15–21
  spin-art, 29–31
  spray-painted, 23–24
  See also wearables
tsuchi dango, 245
Twitter, 265

U

ultraviolet color-changing sun bracelet, 38–39
Unsplash, 76
Urban Buzz, 224

V

video projects, 74–76
voltage, 77

W

water bottle maracas, 153
water gun battles, 254–255
WAV files, 167
wearables, 37
  3D-printed backpack tags, 62–65
  cosplay projects, 55–61
  LED crafts, 40–41
  light-up proximity friendship necklace, 42–47
  light-up sunglasses, 50–54
  stamped leather bracelet, 48–49
  ultraviolet color-changing sun bracelet, 38–39
  See also T-shirts
weather stations, 214
  anemometer, 218–219
  barometer, 217
  high-tech, 220–222
  kits, 220
  low-tech, 214–219
  with an open-source controller, 221–222
  rain gauge, 214–215
  thermometer, 216
  wind vane, 219
Weather Underground weather station, 220
websites for makers, 4, 5
WeVideo, 74
Willeford, Tom, 104
wind vane, 219

Y

YouTube, 76

Z

Zombie Tag, 42